George M. Mowbray

Tri-Nitro-Glycerin

George M. Mowbray

Tri-Nitro-Glycerin

ISBN/EAN: 9783337337919

Printed in Europe, USA, Canada, Australia, Japan

Cover: Foto ©berggeist007 / pixelio.de

More available books at **www.hansebooks.com**

TRI-NITRO-GLYCERIN,

AS APPLIED IN THE

Hoosac Tunnel, Submarine Blasting,

ETC., ETC., ETC.

BY

GEO. M. MOWBRAY,

NORTH ADAMS, MASS.

1872.

NORTH ADAMS:
JAMES T. ROBINSON & SON, PRINTERS AND BINDERS, TRANSCRIPT OFFICE,
Transcript Building, Bank Street.
1872.

Entered according to Act of Congress, in the year 1874, by
GEORGE M. MOWBRAY,
In the Clerk's Office of the District Court of the District of Massachusetts.

DEDICATION.

To WALTER SHANLY, M. P.

Indebted to you for the resources which have enabled me to investigate the properties of Nitro-Glycerin, and render its manufacture a commercial success, permit me to dedicate the following pages in token of my appreciation of the indomitable energy, admirable organization, integrity of purpose, and engineering talent which have rescued the Hoosac Tunnel from the mire of politics and rendered it an engineering success; notwithstanding extraordinary impediments of flood, water fissures, strikes, jealousy and indifference on the part of those chiefly interested, that must have been most disheartening to your mind, and challenged a resolution and resources seldom combined with the abilities you have shewn in this work. Our relations during the past three years having been without a ripple, render this, my simple duty, an agreeable task.

GEO. M. MOWBRAY.

PREFACE.

A paper read by request at the Albany Institute, was the germ of the following pages; its publication in this form, I considered would furnish engineers, contractors and railroad directors, who occasionally apply to me for particulars as to the use of Nitro-Glycerin in the Hoosac Tunnel, with detailed information impossible to condense in a business letter. Hurriedly composed during the spare hours of a manufacture involving grave responsibility, the writer weighted with the additional task of defeating an attempt to monopolize the use (not the manufacture) of Nitro-Glycerin throughout the United States, whilst the subject itself, "Explosives, and firing mines by Electricity," constantly demanded experimental research, this work has not the arrangement nor the completeness I could desire; but the author hopes it will create a more favorable regard in the public mind, towards the most powerful blasting agent known, by correcting errors in respect to its properties, and the casualties attending its use; and assist miners and contractors to a more intelligent acquaintance with some of the materials the present advanced state of engineering progress has brought into practical use.

GEO. M. MOWBRAY.

North Adams, Mass., June 1st, 1872.

CONTENTS.

CHAPTER I.

Nitro-Glycerin — Introduction of the explosive in New York, San Francisco, Lake Superior, and the Hoosac Tunnel, Massachusetts; Accidents; Reports of Engineers Thos. A. Doane, W. P. Granger and B. D. Frost, of the Manufacturer; Miners' statement.

CHAPTER II.

Submarine Blasting — Erie Harbor — Dimon's Reef, New York — Coenties Reef, N. Y. — Oil Wells, Penn.

CHAPTER III.

Nitro-Glycerin considered in its chemical details.

CHAPTER IV.

Electricity in blasting operations.

CHAPTER V.

The Tri-Nitro-Glycerin manufactured at the Hoosac Tunnel — How Tri-Nitro-Glycerin is made — How stored — How Gutta-Percha is purified — How the Exploders are manufactured.

CHAPTER VI.

Explosive mixtures.

CHAPTER VII.

Nitro-Glycerin patents and litigation.

CHAPTER VIII.

Hoosac Tunnel — Drilling by machine — Blasting with Powder — Nitro-Glycerin.

DIRECTIONS FOR HANDLING AND USING TRI-NITRO-GLYCERIN.

APPENDIX.

A. Memoranda for Contractors.
B. Over-sensitive Exploders.
C. Professor Abel on effects of initial explosion on explosives.
D. Car freighted with 4,800 lbs. Nitro-Glycerin off the track.
E. Accidents at the Hoosac Tunnel.

ILLUSTRATIONS.

		PAGE.
I.	Vignette.	
II.	Drilling machine at heading, a photograph taken in Tunnel by Magnesium light, 7,760 feet from West Portal.	
III.	Stereoscopic view. Twelve cans after an explosion,	18
IV.	" " West End, Hoosac Tunnel,	28
V.	" " East End, Hoosac Tunnel,	39
VI.	" " Nitro-Glycerin factory,	43
VII.	" " " " interior of converting room,	46
VIII.	" " Central shaft, Hoosac Tunnel,	50
IX.	Miners ascending " " " "	58
X.	Bursting of can, whilst conveying Nitro-Glycerin, Hoosac Tunnel,	66
XI.	Sinking Central Shaft, Hoosac Tunnel,	74
XII.	Profile of the Hoosac Mountain, shewing progress January 1, 1872,	80
XIII.	"Stopeing out" enlargement, East End,	85
XIV.	Driving bench work and dumping from heading, West End,	90

(Photographs taken by L. Daft, operating for Messrs. Thompson & Co., of Albany, the drawings by Assistant Engineers C. O. Wederkinch and G. Lunt, the wood-cuts by Andrew & Son, Boston.)

CHAPTER I.

Nitro-Glycerin--Introduction of the explosive in New York, San Francisco, Lake Superior, and the Hoosac Tunnel, Massachusetts. Accidents, Reports of Engineers Thos. A. Doane, W. P. Granger and B. D. Frost, of the Manufacturer, Miners' statement.

The city of New York was startled one fine Sunday morning (1865) by an explosion in Greenwich Street, opposite the Wyoming Hotel, the windows of every house within one hundred yards of the entrance to the Wyoming Hotel were shattered, pedestrians were thrown down, and the pavement broken up. A few minutes previous to the explosion, one of the guests in the hotel had been engaged polishing his boots; for this purpose he had drawn from under the counter of the hotel office a small box, on which he had rested his foot; noticing a reddish vapor emanating from there, he drew the attention of the hotel clerk to it, who taking the box in his hands made his way to the front door and threw it into the gutter, whereupon explosion instantly followed.

An investigation of the circumstances connected with the storage of this box, developed the following facts: Some time previously a passenger from Germany who had occupied a room at the hotel, being unsuccessful in obtaining employment had left it as security for his board, stating that it was Glonoin Oil, a new material that had been used in Germany for blasting purposes with great success, that he, the passenger, had been intrusted with an agency for introducing the same to miners and others, but had failed to get it introduced into use; undoubtedly the box contained Nitro-Glycerin, manufactured by the Nobel Brothers, who had a manufactory where this explosive was compounded, at Hamburgh.

In the early part of the year 1866 this substance was again a prominent subject of discussion, owing to an explosion which was attended with the burning and ultimate destruction of the steamer "European," one of the West India mail packets, while she was lying at the railway wharf of Colon or Aspinwall, on the Atlantic side of the isthmus of Panama. Knowing that Nitro-Glycerin was on board under the name of "glonvene" or "glonoin oil," on its way to the gold mining districts of the North American Pacific States, as an explosive or blasting agent, it was concluded that the explosion was due to this substance. Unfortunately, forty-seven persons were either killed at the time of the explosion or died shortly afterward from the injuries they sustained. Immediately succeeding this accident another explosion occurred in the office of Wells, Fargo & Co., in San Francisco, by which eight persons lost their lives. The damages by the explosion on board the "European" were estimated at one million dollars, for the vessel, built of iron and of unusual strength, was destroyed, and the pier with an upper railroad track for unloading cargo, and warehouses for storing freight, were completely wrecked. The San Francisco explosion involved a further loss of a quarter million dollars.

In all the above cases the Nitro-Glycerin manufactured at Hamburgh reached New York safely; in the Wyoming Hotel explosion it had been lying in the hotel several weeks, in the Aspinwall catastrophe it had been transported over the Isthmus and reshipped by steamer as express freight by Wells, Fargo & Co., to San Francisco, and carted to their office in Montgomery Street before the explosion occurred. It subsequently transpired that the immediate cause of the explosion at Aspinwall was, a case slipping from the slings whilst being hoisted out of the hold of the vessel; in San Francisco, the circumstances as detailed to the writer, were as follows: a man passing by Wells, Fargo & Co.' osffice heard one of the employe's address a gentleman riding past on horseback, saying, "Doctor, we have got a case of glonoin oil and it seems to be smoking, I wish you would step in and advise us what had better be done with it;" the doctor (Hill) dismounted, requesting a passer by to take charge of his horse and walk it up and down the block, the animal being too high spirited to stand without an attendant; scarcely had the person in charge gone a block from the office when the explosion occurred. It can only be inferred that in breaking open the case to discover the cause of leakage of red fumes, the Nitro-Glycerin was exploded. I have since ascertained from the New York consignee of this parcel of Nitro-Glycerin, (Messrs. Nobel's agent) that after the shipment to Panama, which was only a part of the consignment from Hamburgh, the agent leaving another portion in warehouse in Tenth Street, New York, proceeded to Lake Superior in the winter season with a part of the same shipment, where, on arrival and opening the cases, he found it had been packed in bottles surrounded with sawdust, and in congealing had burst the bottles, a portion of the Nitro-Glycerin being found solid in the neck of the bottle. This therefore, if correctly reported, would go to prove the Nobel Nitro-Glycerin expands during congelation.* What had been bottles containing Nitro-Glycerin were now fragments of broken glass, whilst the Nitro-Glycerin itself, owing to the extremely cold temperature of a Lake Superior winter, was found in solid mass of the exact mould of the bottle that had contained it. Upon discovering this condition of the cases and their contents the consignee at Lake Superior telegraphed to his correspondent in New York: "Direct Messrs. Bandmann to throw the cases of Nitro-Glycerin, shipped to them, overboard on arrival." Probably in the belief that the temperature of the upper lakes was the cause of the broken bottles and that the warmer temperature of the tropics and San Francisco did not apply, this advice was neglected.

Reflecting as a chemist upon these explosions, that here was a compound made at Hamburgh, carted to the wharf, loaded on board steamer by the stevedores, voyaging to London, reshipped to Panama, the express portion of it forwarded across the Isthmus by railway, thence lightered to and loaded upon the steamer,

---

*This property distinguishes it from the Mowbray Tri-Nitro-Glycerin. the latter contracting about one-twelfth of its bulk in congealing; further, the Nobel patents claim a preparation which congeals at 55° F., whereas the Mowbray Tri-Nitro Glycerin congeals at 45° F. No further evidence is necessary to prove that a real difference of component parts exists between the two preparations.

bearing twelve days' voyage to San Francisco, where on arrival it is taken to the express office, previous to being forwarded to the mines; now how did it happen, since there is no effect without a cause, after all this handling that an explosion took place? Determined to solve this problem, I undertook the preparation and qualitative examination of Nitro-Glycerin. Residing at that time at Titusville in the oil region of Pennsylvania, where the disastrous results of speculations in oil territory during the previous year, compelled most of us to "masterly inactivity," I had the leisure, whilst my curiosity was piqued to discover, the apparently anomalous properties which this explosive seemed to present, and in 1866, after maturing the process patented April 7, 1868, I inserted a brief advertisement in the Scientific American, offering to manufacture Nitro-Glycerin on a large scale for miners and others. In 1866, I received a communication from Thomas A. Doane, Esq., chief engineer of the Hoosac Tunnel, who was keenly alive to the necessity of more efficient means for driving that work. I extract from his annual report to the Commissioners of the Troy and Greenfield Railroad and Hoosac Tunnel, James M. Shute, Alvah Crocker and Charles Hudson, dated Dec. 19, 1866, and having reference to the work of the current year, as follows:

"Page 21. It has been my continual desire since entering upon this work to learn how to fire several charges at the same time. This I hoped to do of Colonel Tal P. Shaffner, but his coming upon our work was so long delayed, it being something more than a year after his first brief visit here, that it began to seem hopeless. Last spring, in making a visit to the Bessemer steel works in Troy, partly in way of business, but more out of curiosity to see and learn something concerning this process of making steel, it was my good fortune to obtain an introduction through Mr. Holley of the steel works, to J. J. Revey of London. Mr. Revey is connected with the gun-cotton works of London, and was acquainted with the most approved methods of simultaneous firing. He very kindly and fully explained to me the process and gave me a description of the electrical machine and fuses necessary, and also afterwards made a visit to our Tunnel. The Commissioners ordered for me two electric machines, four thousand fuses, and several miles of conducting and connecting wire. These were several months in transit and before their arrival Colonel Shaffner came with his material. His machine for exploding was Wheatstone's magneto-electric exploder, and by it and his system of connecting wires it was found impossible to fire more than about five charges at once, and these not simultaneously. This of course was far from satisfactory. Shortly after, the ebonite (or Austrian pattern) machines with the Abel fuses ordered for me, arrived, and we very soon learned how to use them both, and have been able to fire at once as many as thirty-one charges.

"While it is important to save the time which can be saved by this process in firing, and to reduce the risk of accident, and to avoid the smoke made by the burning of the common fuse, it is much more important to the progress that simultaneity of firing be secured. If charges in adjoining holes can be fired as though but one charge, then they help each other and much more rock will be torn away. The whole top may be thrown down or the bottom brought up by proper arrangement of holes, and by means of a ring of converging holes the center may be dragged out. The passage of the electric spark through one system of wires occupies practically no appreciable time, while through several systems it may. If the charges in adjoining holes are fired with the interval of an instant, it may just as well be a week so far as the tearing of the rock is concerned."

"The number of fuses obtained was so small that their influence upon progress is hardly appreciable, except possibly at the Central Shaft.

"Under the direction of Colonel Shaffner, experiments have been tried at the West Shaft with Nitro-Glycerin. The article used was imported from Europe, and much time was consumed in ordering, shipping, and passing it through the custom house. In these experiments Colonel Shaffner has been eminently successful. No accident has resulted, and indeed there seems to be comparatively little risk if the article is good and ordinary care is taken in its use.

"The Glycerin will occasion to some persons, if they are exposed to it in a particular manner, a headache* for an hour or two, while others are not thus affected. Our men have made

---

*This effect has never been produced by the Tri-Nitro-Glycerin ("Mowbray's") and is another and very emphatic proof of the difference between the two preparations.

very little complaint in this respect, and indeed there has been no difficulty experienced in introducing this new and powerful explosive among men who never before have used anything but powder.

"It was some time ago demonstrated by experiment, that double progress could be made with Glycerin over that made with powder at less cost. This is a wonderful achievement and its effect upon the prospect of this work, in regard to its early completion at reasonable cost cannot but be good. It is true that the experiment was limited to a shorter time by reason of the small supply of electrical fuses and Nitro-Glycerin than could have been wished, and that my views may upon further experience be modified or changed even, but with what information I now have there is no room to doubt its fitness for our purpose. It is the testimony of all who have seen our work, including Mr. Revey, George Berkeley of London, C. E., Dr. Ehrhardt of London, Colonel Shaffner, and others familiar with tunnelling, that while our rock is not in general harder to drill than many others, it is most persistently tough. That is, the number of charges we fire, if they could be in granite or lime or in any brittle stone, would bring out two or three times more of debris than now. It is therefore necessary that we should have the quickest explosive to get the best result. As preparations of mercury are not to be thought of on account of their danger, we take Nitro-Glycerin as being next in power, while it is comparatively safe. Whenever its extensive use shall be concluded upon it will be necessary to secure the services of some scientific person expert in handling it, that some antidote against headache may be discovered, and that the risk may be reduced to the lowest possible point. Bulk for bulk, which is the only useful comparison to be made here, Nitro-Glycerin is eight times more powerful than common powder."

In same report, page 64, the consulting engineer, Benj. H. Latrobe, states: "In the east heading of the West Shaft experiments with Nitro-Glycerin as an explosive were made with highly favorable results, as reported by the chief engineer who states, the forward progress in the heading proper (six by fifteen in section) as doubled, and in the heading enlargement (to ten and a half and fifteen) as trebled by this new agent when compared with gunpowder. He also reports $10.20 per cubic yard saved in the heading, and $3.64 in the enlargement, on a similar comparison with gunpowder, results certainly of the most encouraging character, and inviting to farther and persevering effort for the safe and successful use of the new explosive."

The Commissioners themselves report—page 6: "The value and economy of Nitro-Glycerin as an explosive seems to have been fully demonstrated and the method of using it with safety to the employees appears to be the only question now undetermined. Its early introduction is very desirable and preparations are making to bring this about whenever it shall appear prudent to do so, since it is believed, on the strength of numerous experiments made in the tunnel at the West End, that by the use of this agent alone, as compared with gunpowder, the time required for completing the work may be greatly reduced."

Between the issuing of the above report and that of 1867, circumstances led to the withdrawal of Mr. Doane from the Tunnel, and Commissioner Hon. Alvah Crocker personally undertook the superintendence of the work. In his report dated January, 1868, the following remarks occur:

"Nitro-Glycerin—experiments as made in the West Shaft as given by Mr. Doane and referred to by Hon. Tappan Wentworth, chairman of the Tunnel Committee of that year, induced early action by the Commission. As long ago as February last I visited New York, and spent several days in endeavoring to ascertain if the article had been made there, or in the vicinity, but to no purpose. Finding subsequently that the railroads refused absolutely to transport it, the matter rested until the first of July, when I addressed George M. Mowbray, Esq., of Titusville, operative chemist, and with the permission of the Commission he was called to North Adams and a contract concluded with him highly advantageous to the Commonwealth. As will appear in the appendix, the public will be gratified to learn that we are on the eve of giving it a fair trial."

On the 29th of October, 1867, the writer arrived in North Adams and I subjoin my report to the superintending commissioner, dated January 11, 1868, and addressed to Hon. Alvah Crocker, Superintendent of Hoosac Tunnel:

"Sir: I avail myself of permission to report progress of the arrangement to introduce Nitro-Glycerin for the purpose of blasting in the Hoosac Tunnel, subject to the conditions imposed by you at an interview held in the engineer's office, during the latter part of October, 1867. These conditions were—

"First. To conduct the operations with a strict regard to the safety of the miners, and to avoid all risks that might endanger the property of the State, connected with the Tunnel

"Second. The outlay of capital for the necessary works to be defrayed at my own cost and expense.

"Third. That the Nitro-Glycerin should be supplied at current market rates, freight added; the State of Massachusetts furnishing a convenient site for the buildings, compressed air, and a supply of water, free of cost, and to give the subscriber a preference in consideration of his erecting the works adjacent to the Tunnel.

"The reasons that led to this arrangement were, that as the rock found in excavating the Tunnel was exceedingly tough, any increased progress or lineal advance per month without any increased expenditure; in other words, diminished cost per lineal foot and quickened advance, seemed possible only by the use of a more effective explosive agent than gunpowder; that in Nitro-Glycerin this greater power existed, and therefore its use was desirable; the problem being convenience of supply, guarding against the possibility of accident, by planning carefully every detail in its use, rigidly enforcing every precaution, a failure in any of these points involving pecuniary loss in outlay for the works by the party undertaking its supply and superintending its use in the Tunnel.

"Agreeing with you in the propriety of these views, I commenced operations on the 30th of October. During the past two months a convenient two-story factory has been erected, and the necessary apparatus set up therein, about 1000 feet south of the west shaft; within twenty feet of this factory, a small dwelling for myself and an experienced assistant, and about 500 feet further south on the extreme line of land owned by the State, a magazine for storing Nitro-Glycerin has been constructed. Inclement weather somewhat retarded these operations, nevertheless, the crude articles used in the manufacture and every appliance to render the labor of making a "chemically pure" Nitro-Glycerin, without danger to those engaged in its manufacture, were completed and in good working order on the 31st of December, 1867.

"The assistance rendered me by the gentlemen superintending the various departments of the tunnel work, materially contributed to this result, and I gratefully acknowledge their uniform courtesy and promptitude in forwarding my undertaking. Your own constant attendance at the engineer's office permitted me almost daily to submit my plans, which therefore met no delay in being subjected to the scrutiny of the engineer in charge, who as promptly reported on them.

"On the 2d of January, 1868, I moved up to the works and on the following day tested the apparatus by manufacturing, and although somewhat delayed by the necessity of drying the plastering in the magazine, and introducing suitable heating apparatus to maintain a moderate temperature during this inclement season, (a neglect of which precaution remotely led to the Bergen accident) yet to-day we have a supply of Nitro-Glycerin, properly and safely stored, ready for use. Samples of this have been duly tested for its explosive force by the engineer in charge and his assistant, giving satisfaction as to its tremendous power, and facility of explosion, with a peculiar fuse and exploder. You may therefore rely on a regular supply as needed, and I submit that a month's consumption be kept on hand, in order that it may free itself from adherent water, which, except other means be used to free it, does not separate for about ten days. Freed from this obstinately adhering moisture, it is safer and more effective for blasting purposes.

"As respects its application to blasting, during the ensuing week the conducting wires will be laid to the east heading (west shaft) and in order to maintain the electrical machine in working order, I have arranged that the act necessary to firing a blast shall be performed in the time-keeper's office, where the air is dry and therefore favorable to exciting the charge of electricity, but the control and the means to signal for a discharge, will be in the Tunnel at a safe distance from the heading. By this arrangement, although requiring more conducting wire, the incessant repairs to a costly and delicate instrument and disappointment and delay attending miss-fires will be avoided, and the drillers will be detained from their labor at each discharge for a less period of time.

"The order of charging and firing is as follows: When the drill-holes have been completed, (say every four hours) signal is made, for the cartridges which are only then taken into the Tunnel, (the Nitro-Glycerin in its containing cartridge in one vessel, the exploders, with priming and connecting wires attached, in another separate vessel.) On arrival at the heading, the miners are dismissed to a safe distance, the drill-holes are then gauged, to be assured they will receive the cartridges; now, and for the first time the exploders are attached to the Nitro-Glycerin cartridges, and immediately passed into the drill-holes, these latter are plugged with a bung, perforated to allow the delicate connecting wires to pass, (thus avoiding cutting the insulation against the rock, and confining the flame;) connection is made beginning with the return wire to the cartridges consecutively, and on to the conducting wire. The operator now retires from the heading some 300 feet towards the shaft where a simple but important apparatus, or break is arranged; he then and there connects his return wire and his conducting wire to two similar wires that lead to the electrical discharge, which duty is performed in the dry, warm room before referred to, and the explosions take place instantaneously.

The above modification is a necessity to avoid the damaging influence of the moisture in the Tunnel, so disturbing in its effect on the machine. I have only to add, that we have under-way apparatus for coating and re-covering damaged insulated wires, an improvement to insure perfect explosion of the Nitro-Glycerin; the manufacture of Abel's priming for fuse, the formula having been published by the inventor; matters of comparatively minor importance, but where so many blasts are daily occurring, involving considerable saving in cost and express charges, and securing a better article when made by the individual for his own actual use, than when made simply for sale, all tending to greater safety and certainty in firing the blasts, ameliorations that have already been submitted to and approved by your engineer in charge; who will doubtless speedily report the actual results of blasting operations. Respectfully,

Geo. M. Mowbray, Operative Chemist.

The following letter from the Engineer in charge to the Commissioners, is interesting, as showing that the Nitro-Glycerin we had made, was superior, and possessed far more valuable properties, than that which had been imported from Hamburg:

NORTH ADAMS, FEB. 18, 1868.

To the Commissioners of the Troy and Greenfield Railroad and Hoosac Tunnel:

GENTLEMEN:—I have to report that yesterday 4 P. M., we exploded eleven cartridges of Nitro-Glycerin in charges of 1-2 lb. each, in open holes without tamping, with entire success. This experiment was made in the East heading of West Shaft. On approaching the heading, the absence of foul gases and smoke was remarkable, the mass of broken rock lay close to the heading, and there was no appearance of any rock thrown to any distance from the heading. Inquiring of the miners if they experienced any headache, elicited the remark they noticed a pleasant smell, but nothing further. This settles the question of its applicability in a close tunnel. I attribute this freedom from the foul gases which we noticed in our experiments a year since, to the evident purity of this Nitro-Glycerin; it differs greatly from all descriptions of the article, and in appearance from that we imported, being a liquid colorless as water, and free from smell or bubbles. That which we imported was a thick, yellow liquid, quite different in appearance from this. I have requested Mr. Mowbray, who manufactures the Nitro-Glycerin, to take charge of the blasting, and informed him that the Commissioners wish him to assume the responsibility of using the Nitro-Glycerin until further orders, or at least until the system of firing is thoroughly organized among the employees.

I enclose his reply, and approve his suggestions, subject to your instructions.

I am very truly truly yours,

W. P. GRANGER, Engineer in charge.

The Commissioners for the year 1868, report as follows:

During the Summer, Glycerin of a very good quality has been manufactured at this point, under the direction of Mr. Mowbray, and has been used for several months in blasting in the tunnel east of the West Shaft. No accident has attended its use. And while its effect in the heading did not meet the expectations of the Commissioners, the result of its operation in the bench below the heading, justifies the belief that with due provision for its economic use, and essential care and attention paid upon its management, it will prove an effective agent in the prosecution of this enterprise.

The Superintending Engineer, Benj. D. Frost, Esq., reports as follows : "The following is a statement of monthly progress.

	Length driven.	Total distance from W. Shaft.
In November, 1867,	33 feet,	1272 feet.
December, 1867,	22 feet,	1294 feet.
January, 1868,	33 feet,	1327 feet.
February, 1868,	35 feet,	1362 feet.
March, 1868,	34 feet,	1396 feet.
April, 1868,	24 feet,	1420 feet.
May, 1868,	26 feet,	1446 feet.
June, 1868, (1)	21 feet,	1467 feet.
July, 1868, (Nitro-Glycerin used)	47 feet,	1514 feet.
August, 1868, "	44 feet,	1558 feet.
September, 1868, " (2)	51 feet,	1609 feet.

"But for the improved methods of working introduced, the advance would have been much less satisfactory than that we are enabled to exhibit above.

"Concerning the employment of Nitro-Glycerin and machine drilling at West Shaft, it is hardly necessary to remark that many difficulties are to be encountered in the training of men to a new service and in successfully employing a new description of fuse and explosive. Some remarks upon our experience in blasting with this compound, will be found in a subsequent portion of this report. Continuous use of machine drills was commenced at the West Shaft in the latter part of June, and of Nitro-Glycerin as an explosive in the month of August. Some delays were necessarily experienced at first, but greatly improved progress was shortly attained. Some previous trials of machine drilling had been made earlier in the present year, but without continuous progress, upon which satisfactory estimates of success might be based. The last workings made, including the month of September, up to the time of suspension, about five-sixths of a working month, attained a linear progress of 51 feet, with six drills only. The machinery provided at West Shaft is only sufficient to supply the pneumatic power for the

---

(1) Preparing for machine drilling.
(2) September 1, to 24, 5-6 month. Rate 61 feet per month.

ordinary working of the above number, to which accordingly we have been necessarily confined.

The two drill carriages employed are larger than those at East End, and are intended to carry five drills each—in all, ten drills working at the breast of the heading. Assuming, as we may safely, that the rate of progress is proportional to the number of drills employed, ten drills would advance 100 feet per month; and with full power provided and further experience to be acquired by the workmen, this and even greater average rates of monthly progress can be made and maintained.

These headings are run at top, i. e., above the excavations hereafter to be made, and of such height, and top outline as to correspond with the roof of the completed tunnel.

Amounts of progress upon this section of the work during present and preceding year are exhibited in the following comparative table:—

West Shaft Section.	Heading and Adit.		Enlargement.	
	Linear Feet.	Cubic Yards.	Linear Feet.	Cubic Yards.
YEAR ENDING				
November 1, 1867,	543	2349	161	2100
November 1, 1868,	1280	4696	82	488

The limited employment of Nitro-Glycerin made previous to August 1st, had been directed to excavations of enlargement, which very nearly resemble open cut work. The experience of the two months, August and September, is all we have that throws direct light upon its value in mining operations, using this phrase in its more limited sense, as applied to advance of heading only. The varying hardness and tenacity of rock and other attendant conditions, make material variations in the progress of separate days or weeks, even in the same drift and with the same means and appliances of working.

For the reasons thus stated, actual records of advance without full knowledge and discussion of all attendant circumstances, and more especially when confined to short periods, must not be held conclusive in regard to the measure of advantage to be derived from its use. We cannot claim that in this short time, full knowledge as to its best possible application has been obtained. Its superiority over the powder ordinarily used in blasting, as demonstrated by our experience may be briefly expressed in the following items:

"1. Less number of holes drilled in proportion to area of face carried forward. Estimated saving 33 per cent.

"2. Greater depth of holes permissable. Average depth of Nitro-Glycerin, 42 inches; for blasting powder, 30 inches.

"3. More complete avail of the full depth of hole drilled. The greatly superior explosive power of the Nitro-Glycerin rarely fails to take out the rock to the full depth of the hole. Powder often comes short of this, and by reason of this loss of useful effect, a large percentage of additional drilling becomes necessary.

"In all the foregoing comparison, I assume it to be understood that simultaneous blasting by electric battery is employed. The great economy of force secured thereby, whenever hard rock may be encountered, is admitted by all conversant with the matter, and since the early part of the Summer, I have continuously employed it in both the headings advancing into the mountain.

"It is hoped and expected that further experience will demonstrate an increase in each of the several items of advantage resulting from Glycerin blasting; and it is only claimed that the best use was made of the short term of experiment afforded, and the most faithful and diligent effort was put forth to attain the best results and greatest benefit therefrom to the Commonwealth.

"It was a source of great disappointment that Professor Mowbray should have been unable sooner to provide a continuous supply of the explosive, and in view of the fact that a small quantity was furnished earlier in the year, it is appropriate to make mention of the obstacles which for a time delayed its further manufacture. The first lot produced was made with imported acids, reaching the actual standard of purity represented. In providing for more extended operations, acids were ordered of American works of the same expressed standard, but these when received, were found so far below requirement, that a separate process of purification became necessary. For this process, retorts of a special pattern not to be procured in market, had to be manufactured, and separate works erected, and in the processes, necessity for which was not foreseen, much delay was unavoidably encountered. I have been fully satisfied throughout of Professor Mowbray's earnest desire fully to meet the expectations of the Commissioners and of the public, and deem it proper to make this general statement of the more important circumstances, unanticipated, and therefore beyond his control, which disappointed his purpose."

I have been thus explicit in narrating the various details connected with the introduction of Nitro-Glycerin at the Hoosac Tunnel, in order that full justice might be done to the gentlemen whose enterprise and authority were necessary to bear up against the prejudices which the three explosions hereinbefore referred to had caused on the public mind. It is now five years since I commenced, and have with slight intermission, continued, to manufacture this explosive, and during this whole period but two accidents have occurred at my works. The first occurred on the 23rd of December, 1870, to my foreman, who I surmised, in the absence of proof, in removing the clinkers from the heater, may have thrown a red hot coal on to the inflammable floor boards of the magazine, moistened with Nitro-Glycerin spilt during three years use, whilst adding fuel to the parlor stove which warmed it. It is a poor consolation that Mr. Velsor, the foreman, who had been engaged with me during the greater part of the past ten years, had finished his day's work and was using the magazine for a bath house, probably on account of its seclusion. Universally respected, thoroughly acquainted with the properties of Nitro-Glycerin, careful and untiring, cool, courageous, and without bravado, his superintendence of the factory where thousands of pounds of this explosive were being handled, and in the course of distribution to different points of the United States, was steadily and quietly overcoming the dread of this powerful blasting agent; accompanying me and aiding in the most difficult operations of submarine blasting, in every case without a shadow of accident, lead to one conclusion, that some slip of the hand, failure of a muscle, started a flame, which in a magazine crowded with receptacles for Nitro-Glycerin no human power could arrest, but which I am satisfied, his courageous sense of duty led him to attempt, and thereby sacrificed his valuable life.

ACCIDENT IN MAGAZINE.

The new magazine had hardly been completed, and stored with Nitro-Glycerin, when on Sunday morning, 6:30 o'clock, March 12, '71, the neighborhood was startled by another explosion of sixteen hundred pounds of Nitro-Glycerin. The cause of this last explosion, was continuous overheating of the magazine. Work at the factory had been suspended for a week, the heating arrangement was now effected by steam, in order to avoid a possibility of actual fire, the weather for several days had been close and muggy,—some parties who had visited the magazine remarked to me afterwards, they had noticed a hot, close air, similar to that experienced on entering the drying room of a print factory, whilst the watchman confessed he had neglected to examine the thermometer, made up his fire under the boiler, and gone to bed. . I had been summoned during the previous week to Washington, taken down with sickness and unable to return home,—the new foreman having been closely at work without any Christmas vacation, owing to the previous accident, availed himself with my permission, (during the suspension of work at the factory) to visit New York. Fortunately this accident involved no damage to life or limb, whilst a very instructive lesson was taught in the following circumstance: within twelve feet of the magazine was a shed, 16x8 containing twelve 50 lb. cans of congealed Nitro-Glycerin ready for shipment. This shed was utterly destroyed, the floor blasted to splinters, the joists rent to fragments, the cans of congealed Nitro-Glycerin driven into the ground, the tin of which they were composed perforated, contorted, battered, and portions of tin and Nitro-Glycerin sliced off but not exploded. Now, this fact proves one of two things, either that the Tri-Nitro-Glycerin made by the Mowbray process, differs from the German Nitro-Glycerin in its properties, or the statements printed in the foreign journals as quoted again and again that Nitro-Glycerin when congealed is more dangerous than when in a fluid state, are erroneous.

The following incident is, to say the least, instructive: during the severe winter of 1867 and 1868, the Deerfield dam became obstructed with ice, and it was important that it should be cleared out without delay; W. P. Granger, Esq., engineer in charge, determined to attempt its removal by a blast of Nitro-Glycerin. In order to appreciate the following details, it must be borne in mind that the current literature of this explosive distinctly assert-

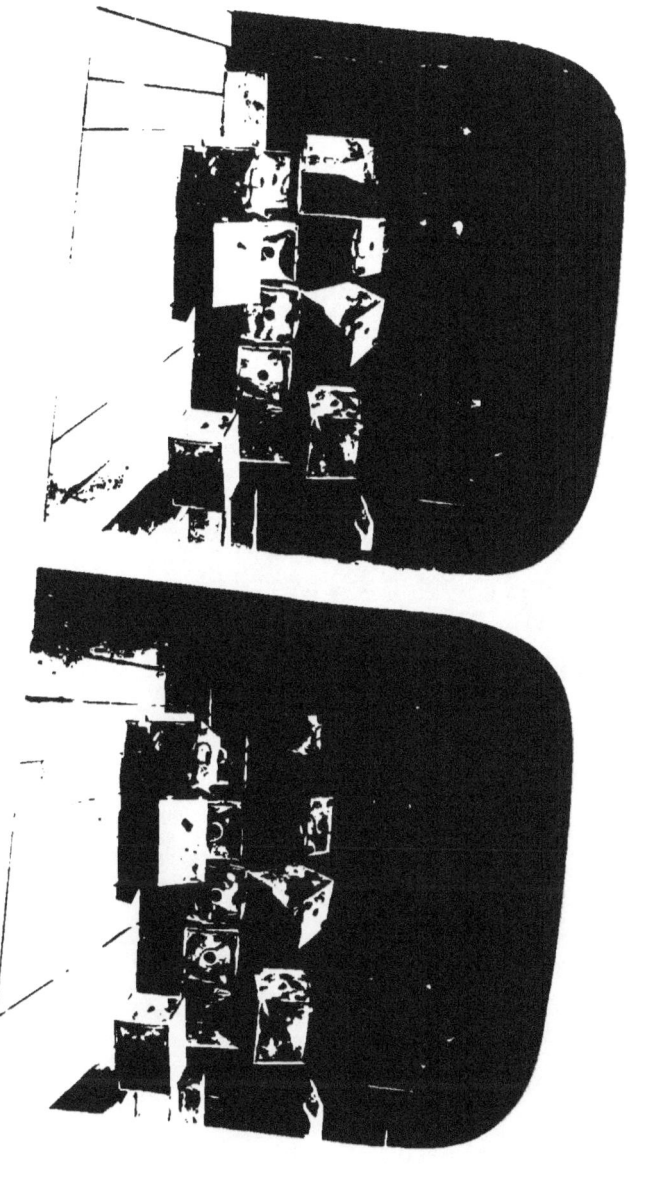

ed that when congealed, the slightest touch or jar was sufficient to explode Nitro-Glycerin. Mr. Granger desired me to prepare for him, ten cartridges, and as he had to carry them in his sleigh from the West end of the Tunnel to the East end or Deerfield dam, a distance of nine miles over the mountain, he requested them to be packed in such a way that they would not be affected by the inclement weather. I therefore caused the Nitro-Glycerin to be warmed up to 90°, warmed the cartridges, and after charging them, packed them in a box with sawdust that had been heated to the same temperature; the box was tied to the back of the sleigh, with a buffalo robe thrown over it; in floundering across the divide where banks, road, hedge and water courses were indistinguishable beneath the drifted snow; horse, sleigh and driver were upset, the box of cartridges got loose, and were spread indiscriminately over the snow; after rectifying this mishap, picking up the various contents of sleigh, and getting ready to start again, it occurred to Mr. Granger to examine his cartridges; his feelings may be imagined when he discovered the Nitro-Glycerin frozen solid; to have left them behind and proceeded to the dam where miners, engineers and laborers were waiting to use this then much dreaded explosive, would never do, so accepting the situation he replaced them in the case, and laying it between his feet proceeded on his way, thinking a heap but saying nothing; arrived, he forthwith attached fuse, exploder, powder and some gun cotton, and inserted the cartridge in the ice; lighting the fuse he retired to a proper distance to watch the explosion; presently a sharp crack indicated that the fuse had done its work, and on proceeding to the hole drilled in the ice, it was found that fragments of the copper cap were imbedded in the solid cylinder of congealed Nitro-Glycerin, which was driven through and out of the tin cartridge into the anchor ice beneath, but not exploded. A second attempt was attended with like results. Foiled in attempting to explode the frozen Nitro-Glycerin, Mr. Granger thawed the contents of another cartridge, attached the fuse and exploder as before; this time the explosion was entirely successful. From that day I have never transported Nitro-Glycerin except in a frozen condition, and to that lesson are we indebted for the safe transmission of more than one hundred and fifty thousand pounds of this explosive, over the roughest roads of New Hampshire, Vermont, Massachusetts, New York, and the coal and oil regions of Pennsylvania, in spring wagons with our own teams.

CHAPTER II.

Submarine Blasting,—Erie Harbor,—Dimon's Reef, N. Y.,—Coenties Reef, N. Y.,—Oil Wells, Penn.

In the winter 1869, 1870, I received a communication from the engineer in charge, Major G. Clinton Gardiner, formerly of the United States Boundary Line Survey, concerning the harbor improvements in Erie, Penn., under W. A. Baldwin, General Superintendent of the Philadelphia and Erie Railroad, with a view to blasting in the harbor of Erie, so as to furnish from 15 to 17 feet of water for vessels laying alongside of their wharves, instead of carrying them (the wharves) into deep water; these operations were entirely successful, and I subjoin the report of Major Gardiner to General Parke, U. S. Engineer Corps, written previous to dredging. The certificates of Mr. Baldwin, Superintendent; F. J. Wilson, Ass't Engineer; Chas. F. Dunbar, contractor for the dredging, follow Major Gardiner's report. These certificates it will be observed, were given after a considerable portion of the rock had been removed by the dredging machine.

LETTER FROM MAJOR G. CLINTON GARDINER TO GENERAL JOHN G. PARKE, Corps of Engineers, Washington City, D. C.

OFFICE OF PHILADELPHIA & ERIE RAILROAD.
Erie Harbor—August 2nd, 1869.

To GENERAL JOHN G. PARKE, Corps of Engineers, U. S. A.

My dear General: Some days ago I received a letter from Mr. Geo. M. Mowbray, who is the patentee of a most valuable improvement in the manufacture of Nitro-Glycerin. He being interested in having his material used in the improvements at Hell Gate, requested me to report upon the experiment in blasting at this place. Being unknown to General Newton, and having no time for a report, I take the liberty of writing to you on the subject.

Since leaving the United States Boundary Survey, I have been employed on the Philadelphia and Erie Railroad, under the direction of the Ass't Gen. Superintendent, Mr. W. A. Baldwin, in the improvement of their dock at this terminus of the road. The water at the end of the main pier and for a short distance inshore, on either side of the pier, is over 14 feet deep, shoaling back to about 6 feet, which we had to deepen to 14 feet. The bottom is a smooth hard surface of shale rock, a portion of which when exposed to the air disintegrates, while other parts are sufficiently hard, and are used for, building purposes. It lies in strata of about eight inches to twelve inches thickness, which we drilled through and blasted during the winter, and are now dredging the rock. The process of drilling was in the primitive style, with hand drills, mostly done through the ice, and the blasting, with powder in cartridges with small tubes reaching to the surface of the water, through which the match was conducted to the powder. Firing however, was afterwards done by dropping a red hot nail down the tube, which was quite an improvement on the match, and gave us almost simultaneous explosions. The holes drilled were 5 feet apart, in rows of 5 feet from each other, and the largest charge of powder used was a canister 2 inches in diameter and 40 inches long. This process having been used to some extent the season before, it was commenced again this last winter, but the work being extended, we thought it advisable to make some improvements in the modus operandi. After a correspondence with different manufacturers of machine drills, we found no one of them ready for business at once, and before we were able to make terms, our primitive style of drilling advanced almost to completion. We sent to Mr. Mowbray who was then in Titusville, Pennsylvania, to try his Nitro-Glycerin, and made an experiment in a square of a little over ten yards, where the rock to be removed was over seven feet deep. The holes were drilled a greater distance apart, but to the same depth as used for powder (15 feet from surface of water). In this square we blasted about 230 square yards of rock, using 50 pounds of Nitro-Glycerin in cartridges fired in rows by electricity, but without a face of rock to work from, such as we had with the powder blast. This would have taken 125 lbs. powder. Upon reaching the place with the dredge, we found the rock completely crumbled, RENDERING DREDGING AS EASY AS THAT OF GRAVEL, and to the depth of seventeen feet, while with the powder blasting we have had trouble, and in two cases had to blast again to obtain fourteen feet of water, and even then have to lift rock measuring ten and twelve cubic feet. Nitro-Glycerin is certainly far superior in its effect, and would have been much cheaper to use in this case. Gunpowder does not blast to the depth of the holes drilled, whilst Nitro-Glycerin tears the rock from the bottom, and here seems to have penetrated three feet beyond. The reason it was not used before, was the difficulty in procuring it. The nearest factory was that of Mr. Mowbray at Titusville, and the local as well as state laws were such that it could not be transported, except by private conveyance, which added to its cost. That used was carried to Corry in Mr. Mowbray's carriage, over a very rough road, and thence by special train to this place. If pure, the danger in the use of Nitro-Glycerin is no greater than that of powder, and the premature explosions that have proved so fatal in many instances, have without doubt been caused by decomposition, which was the result of imperfect manufacture. If regularly manufactured, accidents will be the result only of inexperience or the neglect of instructions from those having experience. In the manufacture, the nitrous vapours that are disengaged at the time of mixing, if not entirely expelled, will make it liable to explosion from any concussion, and from Mr. Mowbray's experience in a number of instances with that manufactured by himself, I should judge his Nitro-Glycerin to be as safe as powder in the hands of experienced persons. It is of a light yellowish color, with pungent aromatic taste, rather sweet than otherwise, and is so poisonous, that in handling, should one allow it to remain on his hands, it would produce intense head ache. It does not explode from the application of flame to its surface, yet will burn, but explodes only from severe concussion, as by the explosion of detonating mixtures and fulminates.

I write to you hoping you will communicate any information my letter may contain to General Newton, as it may serve Mr. Mowbray, who I think has made a great improvement in the manufacture of Nitro-Glycerin, and as he gives it his personal attention, I have no doubt it is superior to any now used.

I was much pleased to receive the report of the blasting in California, and should interesting professional papers be published by the Bureau, let me beg you will remember

Your sincere friend,

G. CLINTON GARDINER.

The experiments above narrated and conducted under the supervision of Major Gardiner, were continued, (on the removal of the Major to the Pennsylvania Central's works at Altoona,) by F. J. Wilson, under General Superintendent Wm. A. Baldwin, and the results expected were entirely fulfilled, as will be seen by the subjoined communications:

SUBMARINE BLASTING WITH NITRO-GLYCERIN; RESULTS AS COMPARED WITH BLASTING-POWDER, IN ERIE HARBOR, MAY, 1870.

Philadelphia and Erie R. R.; Pennsylvania R. R. Co., Lessee.

Office of the General Superintendent,
ERIE, PENN., May 19th, '70.

To GEO. M. MOWBRAY,

North Adams, Mass.,

Dear Sir: The comparative values of the two materials, Gun-Powder and Nitro-Glycerin, as to results and actual cost for blasting in the harbor at Erie, cannot be positively obtained until the dredging is finished; when this year's operations with Nitro-Glycerin, can be compared with that of last year done with powder. The prospects thus far are so favorable, however, I regret that the use of Nitro-Glycerin was not adopted last year.

On the completion of the work I shall be pleased to furnish

you with statements of comparative results, feeling confident they will prove a more full satisfactory and valuable endorsement of your Nitro-Glycerin for submarine use, than any theoretically based opinion can be.

I enclose you copy of reports of Mr. F. J. Wilson, Engineer in charge of Erie Harbor Works, and of Mr. Dunbar, contractor for dredging, which will give you an idea of the economical results to us from the use of your Nitro-Glycerin.

Yours truly,
WM. A. BALDWIN, Gen'l Supt.

ERIE, Penn., May 16th, 1870.

WM. A. BALDWIN, Esq.,
Gen'l Supt. P. and E. Railroad.

Dear Sir: Below please find a statement of comparative cost of drilling and blasting where Nitro-Glycerin is used. The 1240 lbs. of Nitro-Glycerin were used over an area of 26,700 sq. feet, with an average depth of rock of about seven and seven-tenths feet, making 11,500 cub. yards of rock measured in the bed.
Cost of drilling and blasting (using Nitro-Glycerin), $5,119 67.
Cost of drilling and blasting (using Powder), 7,475 73.
Difference of cost in favor of Nitro-Glycerin, 2,356 06.

The difference in favor of Nitro-Glycerin in dredging and in time saved is not taken into consideration in the above (see Capt. Dunbar's letter).

Very respectfully,
F. J. WILSON, Ass't Engineer.

ERIE, May 18th, 1870.

To W. A. BALDWIN, Esq.,
Gen'l. Supt. P. and E. Railroad,

Dear Sir: In reply to your inquiry as to the relative difference in dredging rock blasted by Nitro-Glycerin and that blasted by Powder, I have no hesitation in saying that I am certain we can dredge twice the number of cubic yards where it is blasted with the Nitro-Glycerin. I think I could speak safely and say three yards to one where the rock is hard. In fact, there are places where we could do nothing with the Powder blasting, when we have no trouble with the Nitro-Glycerin.

   Truly yours,
    Chas. F. Dunbar,
     Firm of Lee & Dunbar.

Result.—Submarine drilling and blasting with Nitro-Glycerin costs 44 ½ cents per cubic yard. Gun-powder costs 66 ¾ cents per cubic yard. Nitro-Glycerin used, one ounce and six-tenths of an ounce per cubic yard of rock removed.

Dimon's Reef, New York Harbor.

General Newton, U. S. Corps of Engineers, who has been entrusted with the expenditure of the annual appropriation for the improvements in New York harbor, having constructed a floating drilling apparatus, with steam power to capstans, four steam derricks, and direct engines to lift the drop-drills, applied to me (1870) first, to enter upon a competitive test, with Nitro-Glycerin as compared with Dualin, and with blasting powder, into which a reel of lightning fuse was inserted, to ensure more perfect and rapid combustion of the powder. These tests were conducted at Hell Gate, under the supervision of Mr. Reitheimer; Mr. H. H. Pratt, with Nitro-Glycerin, on my behalf; Mr. Ditmar with Dualin, and Mr. Gomez, for the powder and lightning fuse blasts, who respectively directed the holes to be drilled, charged them, and fired the several charges. The results were decisive of the superiority of Nitro-Glycerin, over both Dualin, and Blasting-Powder, even when assisted by a coil of lightning or fulminating fuse, inserted in the powder. Two points were elicited, as reported by my operator; first the Nitro-Glycerin tore out more work, invariably reaching to the bottom, and sometimes beyond the bottom of the drill-hole, whilst its explosion was so instantaneous it did not cause leakage in the roof, as with Dualin. Thereupon I was invited by General Newton, to arrange operations for blasting at Dimon's Reef, between the Staten Island Ferry and Governor's Island. Eight holes had been drilled in a circle of twenty feet diameter, with a ninth or central hole, thus leaving an average of eight feet of rock between each drill hole. Finding that the drilled holes were shaped like an inverted cone, owing to the omission of the reamer; that is, whilst the drill, jars, sinker bar, cable and cable clutch of the Pennsylvania oil wells, had been used, the provision for remedying the effect of the worn edges of the drill, had been overlooked, and thus a very disadvantageous form of hole, viz: funnel shaped, was the result, necessitating the use of a cartridge, whose diameter must not exceed that of the smallest, which in this case was the lowest part of the drilled hole. The irregularity, and jagged edge of these unreamed holes, had also to be guarded against, lest the friction of any Nitro-Glycerin moistening the outside of the cartridge, might cause a casualty. I therefore determined, until better drilling could be secured, to use $2\frac{1}{4}$ inch two-ply rubber hose for cartridges, a material by no means desirable, because it afforded a cushion between the rock and the blast, but it became a necessity from the funnel shaped drill holes, when providing against the risk of premature explosion. The holes being $4\frac{1}{2}$ inches in diameter at the upper part, and barely 3 inches at the bottom; the cartridge made of rubber hose, being uniform throughout, containing a column of liquid Nitro-Glycerin, $2\frac{1}{4}$ inches in diameter only, and 6 feet long; at the upper part of the holes there was an intervening cushion of water and hose, over 1 inch thick; and at the lower part, a cushion of $\frac{3}{8}$ inch of hose. This should have been avoided, and I have mentioned these details as a caution to future operators, who desire the full explosive force of Nitro-Glycerin.

The depth of water at or during high-tide, is about twenty-two feet, and at low tide, fourteen to fifteen feet, the tide running four miles an hour with an amount of silty matter, drainage of N. Y. City sewerage, rendering it impossible for a diver to distinguish objects one foot from his helmet. Under these circumstances plugs have to be inserted in the several holes, each plug attached to the other by a rope, so as to enable the diver to guide himself from one hole to the other. Owing to various interfering circumstances the holes were only ready for blasting on the 16th of December, 1870; and the second day after arrival in New York, accompanied with three assistants, I proceeded to the work; there was a stiff wind blowing from the northwest, which, meeting the tide, caused a chopping sea; the weather was cold as shown by the crust of ice attached to the scow. The frozen Nitro-Glycerin was thawed out by hot water obtained from the steam boiler on board the scow.

These cartridges were lowered to the diver with the connecting wire, fuse, and exploder attached, one after the other, occupying twenty minutes; two of the holes being too small to allow the cartridge to be fully inserted, these projected, one about eighteen inches, the other one foot above the surface of the holes; the diver, moreover, became entangled in the wires and in order to extricate him, it was necessary twice to haul him to the surface, after which considerable time was occupied in moving the scow from over the site of the intended explosion, before the order could be given to fire. The amount of Nitro-Glycerin used to fill the nine cartridges, was one hundred and thirty-four pounds. On the order being given, the charge was successfully fired. Similar charges of nine cartridges, with more perfect holes and a heavier charge were fired three weeks afterwards.

NITRO-GLYCERIN TORPEDOES IN OIL WELLS.—The Legislature of Massachusetts having resolved to place the further construction of the Hoosac Tunnel under contract, pending the transfer from October, 1868, to April, 1869, from State management to the present contractors, Messrs. F. Shanly & Co. I proceeded to the Oil Region, and there verified the fact that Nitro-Glycerin, properly exploded, i. e., the charge completely exploded, was more efficient in causing an increased yield of oil when applied to wells ceasing or diminishing their yield, than any other material. Ehrhardt's powder, Oriental powder, and ordinary blasting powder, had been used very generally, and Nitro-Glycerin had been alleged to have been used, but the results were unsatisfactory; as soon however, as we started a Nitro-Glycerin factory at Titusville, and inserted charges varying from six pounds to fifty pounds, the results were so advantageous to the well owners, that none others would be used, while Nitro-Glycerin could be obtained. The first explosion was in D. Crossley's well on the Weed farm, a charge of six pounds having been inserted, and fired. The well whose previous best yield had only amounted to six barrels per day, increased forthwith to one hundred and twenty barrels of petroleum per day, and settled down to forty barrels per day, which were obtained daily for nearly a year. On the road to Enterprise at the McKinney & Prior well, the explosion of six pounds of Nitro-Glycerin invariably started the well to flow at the rate of about one hundred barrels in twenty-four hours. At the Crocker wells on the Weed farm, the increase after an explosion of Nitro-Glycerin was usually from ten barrels to one hundred and twenty. After a charge of Nitro-Glycerin in an oil well, the yield generally rises to the highest point it has ever attained, and thence gradually diminishes therefrom, apparently owing to an accumulation of paraffine deposited in the interstices of the walls of the well. This has led to the pouring down the well, benzine, and pumping same out with the oil, and is another form of recuperating the yield of oil. As the process of increasing the production of Petroleum in oil wells, by means of the explosion of gunpowder or its equivalent, substantially as described in the specification of E. O. L. Roberts, ante-dated May 20, 1866, was claimed by the patentee to cover the use of Nitro-Glycerin and every known or hereafter to be invented method of effecting an explosion in an oil well, and as the case has hereto been presented in the courts, this claim has been sustained.

When, therefore, the contractors commenced operations on their work at the Tunnel, I resumed my manufacture of Nitro-Glycerin for that work, leaving the oil region, where the oil operators and producers have since been incessantly litigating the validity of the Roberts patents above referred to, with, however, up to the present date, indifferent success. The average of greatly increased production in exhausted wells, so far as my experience extended, during four months at one hundred wells, was that 80 per cent. were benefited, and in about 20 per cent. no marked results were obtained. When the question as to whether this form of blasting, viz: in oil wells, is patentable has been decided, it will be time to renew the careful application of Nitro-Glycerin in oil wells, but at present, the careless handling, the pursuit of wealth regardless of the lives of the employed, and the unscrupulous assertion prevalent among those interested in the patent referred to, is depriving the oil producers of a valuable agent. Since, however, the present yield of oil is ample for the consumption, this, so far as the public is concerned, is of less moment than it is to the producers, who, by the time economical and useful blasting in oil wells is needed to bring up the yield to the ever increasing demand, will have finally disposed of this patent litigation.

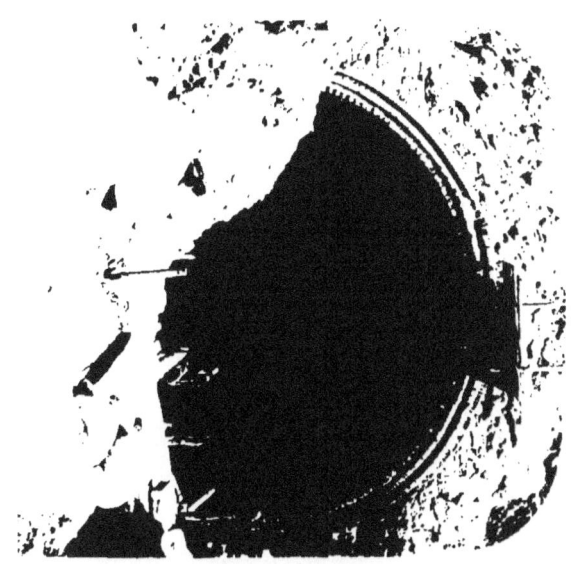
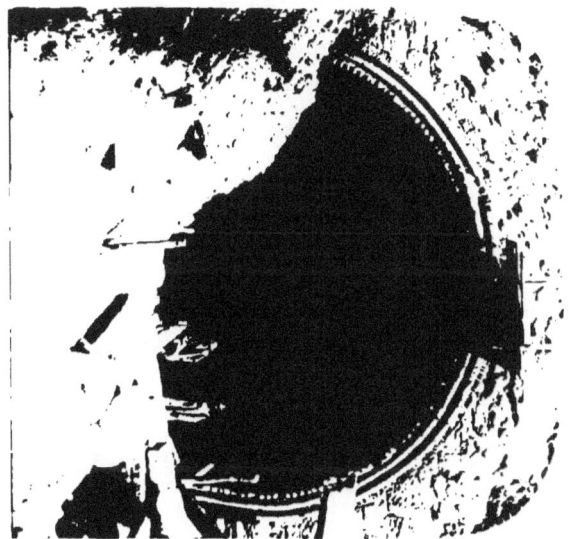

CHAPTER III.

Nitro-Glycerin Considered in its Chemical Details.

GLYCERIN, the base of Nitro-Glycerin, is produced from most of the fixed oils and solid fats by the process of saponification, that is, by treating these fatty bodies with an alkali, or other metallic oxide, in presence of water, or with water itself at a high temperature. For many years the Glycerin of commerce was produced from olive oil, by boiling, in the presence of water, litharge, which yielded the well known lead plaster or diachylon, and a sweetish liquid, which by evaporation of the water, was found to be Glycerin; thus procured, however, it was apt to be contaminated with lead, and therefore very objectionable for medical purposes. The sources whence it is now procured, are, the alkaline mother liquor of the soap works, when the soap is separated by common salt: also the residue of the manufacture of stearic acid for candles, by heating neutral fats with water or with steam, (Tilghmann's process): and the action of muriatic acid on castor oil. It is apt to be contaminated with sulphuric acid, oxalic acid, lead, and more generally with uncrystallizable sugars. The demand has vastly increased of late years for medical purposes, elastic sponge, and retaining moisture in tobacco, print works, as a preserving agent, and for floating compasses, etc., etc.

The following are the synonyms of Nitro-Glycerin; Nitrate of Oxide of Lipyl, (BERZELIUS); Glonoin, Mono-Nitro-Glycerin, Di-Nitro-Glycerin, Tri-Nitro-Glycerin, (LIECKE)—Symbol, ($C^6 H^5$,) O^3, $3NO^5$; (Hydrogen $= 1$, Oxygen $= 8$,) the equivalent or atomic weight is 147.

Pure Nitro-Glycerin is nearly colorless; usually, however, owing to coloring matter contained in the Glycerin used in its manufacture, it is of a light yellow-tinted color, oily, without odor, but having an aromatic taste, Sp. Gr. 1.6 at 60° F., very insoluble in water; mixes with alcohol (one part to four parts) and ether; it separates from the alcoholic solution by the addition of water. A vinous taste is perceptible to the tongue, the maxillary glands are stimulated, and in a few minutes the individual who has tasted it from a pin's point for the first time, is conscious of a persistent, throbbing headache. Slightly touching it with the hands produces a like effect; after a few days of frequent handling, however, the system becomes less susceptible to these effects, and workmen constantly employed in its manufacture are not affected by it. It is poisonous, a small quantity being sufficient to kill a dog, (SOBRERO). It decomposes at 320° F., giving out red vapors, and explodes at a higher temperature, or by concussion or percussion, crashing the containing vessel; it ignites by flame, and burns without explosion, yielding a light ethereal flame of considerable volume.

Pure Nitro-Glycerin may be kept for a year unchanged, (De Vrij). The writer has exposed it to frost, sun and rain, for three years, and found it unchanged. Unless perfectly pure, however, it rapidly changes, becoming of an orange yellow color, evolving fumes, and seems to become a wholly differing compound, being difficult, when thus changed, to congeal, except at a much lower temperature than 45° F., and is more readily exploded.

It very easily decomposes by drying in a warm room with rarefied air, (WILLIAMSON).

It is instantly decomposed when dissolved in alcohol, by adding an alcoholic solution of caustic potash, the reaction being so violent as to eject the mixture from the test tube.

Nitro-Glycerin in contact with the following salts: nitrates of lime, cobalt, soda, barytes and potash; chlorides of calcium, of barium; perchloride of iron, carbonate of lime, sulphates of potash, lime and soda, was found unchanged after a year's exposure.

INCOMPATIBLES: nitrate of silver precipitates black oxide of silver; nitrate of copper gives a precipitate of peroxide of copper, the Nitro-Glycerin remaining, however, bright and apparently unchanged. In a solution of nitrate of mercury, there appears a white film, a bubble of protoxide of azote, apparently adherent to the Nitro-Glycerin. Muriate of ammonia seems to divide the Nitro-Glycerin into two liquids, a light film supernatant, and the heavier liquid subjacent. The action of chloride of mercury (calomel) is but very slight. Protochloride of tin forms a precipitate of peroxide of tin, the residuary Nitro-Glycerin reflecting light powerfully, and as brightly as a diamond. Bichromate of potash is partly reduced to chromate. Sulphate of copper forms a very slight precipitate of oxide of copper, with apparently no change in the residuary Nitro-Glycerin. Sulphate of iron decomposes it, giving a voluminous precipitate, with evolution of nitrous fumes. Sulphuret of ammonia decomposes it, with precipitation of sulphur. Acetate of lead, chlorine water, ferridcyanide of potassium, cyanide of potassium, sulphocyanide of potassium, and of mercury, nitroprusside of sodium decompose it, also the sulphurets of iron, and potassium.

The action of tin, iron, and lead, slowly decomposing the Nitro-Glycerin, especially in the presence of an acid, indicates that metals having an affinity for oxygen, are the most active in promoting decomposition, evolving at the same time nitrous fumes, or protoxide of nitrogen, whilst the residuary Nitro-Glycerin does not seem to be affected; with sulphuretted hydrogen, as with sulphuret of sodium, potassium and ammonium, the action is prompt, and if these reagents be added in sufficient quantity, the Nitro-Glycerin is wholly decomposed, sulphur being precipitated.

Ascagne Sobrero, the discoverer of Nitro-Glycerin, says: it may be prepared by slowly introducing syrupy Glycerin into a mixture of two volumes concentrated sulphuric acid to one volume of nitric acid, Sp. Gr. 1.4, dropping it in and rapidly cooling. It seems to dissolve in this mixture without any noticeable reaction, and by pouring it into water, the so formed Nitro-Glycerin separates from it. It is then washed several times in water, is next dissolved in ether, and after evaporation (dangerous work this) is finally purified over sulphuric acid.

De Vrij recommends dissolving 100 grammes of Glycerin Sp. Gr. 1.262 in 200 c.c. of hydrated nitric acid cooled to 14° F., taking care that the mixture never exceeds in temperature 32° F. When a homogeneous mixture has been obtained, 200 c.c. of strong sulphuric acid are added very gradually, taking especial care that the temperature of this mixture never rises above 32°

F. The oily Nitro-Glycerin which floats on the surface is separated by a tap-funnel from the acid liquid (which yields more Nitro-Glycerin on being diluted with water) and is now dissolved in the smallest possible quantity of ether; this solution is shaken with water, until the water no longer reddens litmus; the ether evaporated (here take care!) and the remaining Nitro-Glycerin heated over the water-bath till its weight remains constant. Merck, of Darmstadt, the eminent operative chemist, found in following De Vrij's method, whilst evaporating the etherial solution, and before the temperature had reached 212°, it was accompanied by a terrible explosion. An accident from the same cause occurred in the laboratory of Dr. E. Von Gorup-Besanez, and we find in "Comptes Rendus" an account of the effects of the explosion of only 10 drops of Nitro-Glycerin, which, by one of the pupils of that chemist, in his laboratory, were put into a small cast-iron saucepan, and heated with a Bunsen gas flame. The effect of the explosion was that the forty-six panes of glass of the windows of the laboratory were smashed to atoms, the saucepan was hurled through a brick wall, the stout iron stand on which the vessel had been placed was partly split, partly spirally twisted, and the tube of the Bunsen burner was split and flattened outwards. Fortunately, none of the three persons present in the laboratory at the time were hurt. When Nitro-Glycerin is caused to fall drop by drop on a thoroughly red-hot iron plate, it burns off as gunpowder would do under the same conditions; but if the iron is not red hot, but yet hot enough to cause the Nitro-Glycerin to boil suddenly, an explosion takes place.

Nitro-Glycerin is decomposed by evaporation, even in vacuo, over sulphuric acid at ordinary temperatures (RAILTON), and when left to itself, frequently undergoes spontaneous decomposition; but when well purified, it may be kept for a long time without alteration (H. WATTS); exhibits different properties, according to the manner in which it is prepared (GLADSTONE).

Liecke in Dingler's Polytechnical Journal, prescribes the following formulæ for manufacturing the three several preparations, Mono-Nitro-Glycerin, Di-Nitro-Glycerin and Tri-Nitro-Glycerin.

Mono-Nitro-Glycerin: Glycerin 100 grammes.
Nitric acid, Sp. Gr. 1.3,200 grammes.

Dissolve the Glycerin in the nitric acid, and then add sulphuric acid 200 cubic centimeters.

The product should be $\left. \begin{array}{l} C^3\ H^5\ O^2\ H \\ NO^4\ H \end{array} \right\} O^4$

Di-Nitro-Glycerin:

Sulphuric acid containing 1 eq. water, two volumes, nitric acid, Sp. Gr. 1.4, one volume; mix the above, lower the temperature to $32°$ F., or below, and drop into it

Glycerin, pure,   one volume.

Prod. $\left. \begin{array}{l} C^3\ H^5\ O^2\ H \\ 2\ NO^4 \end{array} \right\} O^4$

Tri-nitro-glycerin:

Sulphuric Acid, 3.5 parts.
Nitrate of Potash, 1 part.

cooled to $0°$ F., produces $KO + 4\ SO^3 + 6\ HO$, from this the concentrated fuming Nitric acid is separated by decantation, and being maintained at $0°$ F,

Glycerin 0.8 parts is very gradually added,

producing $\left. \begin{array}{l} C^3\ H^5\ O^2\ NO^4 \\ 2\ NO^4 \end{array} \right\} O^4$

From the above extracts of several of the most eminent chemists of the present day, the reader will glean, that in order to prepare this explosive, of uniform quality, invariable in composition, free from water, or any other impurity, it is not merely necessary to buy the best materials, but to have at command the means of verifying their purity before attempting its manufacture.

These points secured, viz: purity and strength of materials, i. e., glycerin free from sugar, fatty acid, saline impurities, and a mixture of Sulphuric Acid with Nitric Acid in due proportion, of due percentage of the respective acids, and not more water therein, nor in the glycerin, at one time of making, than another; the next point to command will be, that in combining the glycerin with the acids, when considerable heat is evolved, the heat thus evolved shall be absorbed rapidly, so as never under any circumstances whatever, to exceed a certain temperature. 'Sobrero names $32°$ F.; otherwise, according to my experience, very differing nitro-glycerin will result from variation of temperature whilst mixing. Such products may be fatal to the miner, although only affecting the manufacturer in a pecuniary sense.

I am led to emphasize these remarks from the fact that prospectuses have been issued to tempt contractors to buy apparatus in the one case, and offering to manufacture on the side of a railroad cutting, if required, in another case, by parties who have no experience in the manufacture, and who start in their new avocation, by deriding the care, outlay and precautions that their competitors have deemed it necessary to make, in order to secure a uniform, certain, and, for mining purposes, perfectly safe explosive; for as the product is to be handed over to the uneducated miner, who cannot estimate the risk he is subjected to even if such a course occurred to him, it does seem to me just and proper, that the controlling engineer, the intelligent contractor, and especially the operating miner who is to handle this explosive, should be advised, that under the term Nitro-Glycerin, very different substances, both as regards explosive force, and liability to spontaneous explosion, do result, unless extraordinary precautions are adopted in the selection of the crude materials, as well as securing uniformly low temperature throughout the process of making. Unless this be done, decomposition sets in and is indicated by the emanation of fumes, by the deepening of the light lemon tint to an orange yellow, and at this point, the miner should decline using it, and require the manufacturer to take his place, and the risks contingent on using it.

Since many of the accidents that have occurred with Nitro-Glycerin, have been traced to leakage from the containing vessel, notably the San Francisco accident, probably the Panama explosion, and undoubtedly the Titusville or Enterprise explosion, besides other cases where it leaked through the bottom of wagon and thence on to the springs, whose hammering caused an explosion, the discovery by Granger, page 19, confirmed by the magazine explosion, page 18, teach the importance of transporting this explosive in a solid state, that is, congealed; there is however another reason; decomposing Nitro-Glycerin will not solidify at 45° F., and the comsumer has a ready and convenient test for the purity of this article, by seeing to it that he invariably purchases the explosive deliverable in a solid form. Another test is, when exploded, in a close tunnel, the fumes or decomposed gases should not inconvenience the miner.—Failing in either of these tests, it may fairly be rejected as an inferior article, or should be used up as speedily as possible, preferably by the manufacturer or his more experienced employees, rather than by a miner who may not be fully aware of the unnecessary risk to which he is exposed in handling impure Nitro-Glycerin.

METHOD OF ANALYSIS.

Walter Crum* describes a method of analysing bodies containing nitric acid, applicable to the nitro-compounds; when nitrate of potash is used, it is previously purified by crystallization, and fused at little more than its melting heat. Nitro-Glycerin, gun-cotton, etc., must be deprived of moisture.

A glass jar eight inches long and an inch and a quarter in diameter, is filled with and inverted over mercury; a single lump of the fused nitrate, weighing about six grains, is let up through the mercury into the inverted jar, and afterwards fifty grains of water. As soon as the nitrate is dissolved, 125 grains of sulphuric acid, ascertained to be free from nitric acid, are added. By the action of the mercury upon the liberated nitric acid, deutoxide of nitrogen soon begins to be evolved, and, usually in about two hours, without the application of heat, the whole of the nitric acid is converted into that gas. Sometimes agitation is necessary, and it is easily performed by giving a jerking horizontal motion to the upper part of the jar. The surface of the sulphuric acid is then marked, and three-fourths of an inch of solution of sulphate of iron recently boiled, let up into the jar. The gas is rapidly absorbed, except a small portion at last, which must be left several hours to the action of the solution, or be well agitated in a smaller tube with a fresh portion of it. No correction of the nitric oxide has to be made for moisture, for the mixture of acid and water employed has no perceptible vapor tension.

In one experiment, 5.40 grains of nitrate of potash yielded 4.975 cubic inches of gas, at 60° F., and barometer 30 inches.

The residue not absorbed by the sulphate of iron, was 0.015 cubic inch, leaving

4.96 cubic inches of nitric oxide $= 1.594$ grains NO^2, and which correspond to 2.869 grains nitric acid, or 53.13 of the nitrate of potash.

---

* Pharmaceutical Transactions, vol. 7, 1848, p. 27, et seq.

Four consecutive experiments yielded
        53.13
        53.14
        53.73
        53.29

Mean 53.32 or leaving out the third experiment. Mean 53.19

The calculated percentage of nitric acid in nitrate of potash, the acid being represented by 6.75, and the potash by 5.8992, is 53.36. THOMSON gives for percentage of nitric acid in nitrate of potash 52.94, and BERZELIUS 53.44.

Salts in powder, which are difficult to pass through mercury without loss, may be enclosed in small glass cylinders. Nitro-Glycerin may be made into pellets with powdered glass, and congealed at 45°, or simply congealed by taking great care it is not partially thawed during manipulation.

Mr. Theron Skeel, of Albany, has furnished me with the following extract from the Engineering Journal of the 17th Nov., 1871, being an explanation of M. L. Hote's method of analysing the gases produced by the explosion of Nitro-Glycerin. He uses Ure's graduated electric eudiometer, made out of a green glass organic analysis tube. Introduce into the apparatus ten centimeters of the gases evolved from water by voltaic electricity, then introduce small globules of thin glass, containing from five to six milligrammes of the explosive; an electric spark being passed through the mixed gases by means of the platina points melted in the upper part of the eudiometer, explodes the gases, breaks the small glass globules and explodes the Nitro-Glycerin. The gases evolved are colorless, and contain a proportion of binoxide of nitrogen. Submitted to the proper absorbents, for moisture, binoxide of nitrogen and carbonic acid, there remains nitrogen. Thus:

1 gramme Nitro-Glycerin gave at temp. 0 Cent.
                        29.7 barom. press., of these gases 284 c.c.

One hundred parts by volume contained
        Carbonic acid,   45.72
        Binoxide of Nitrogen, 20.36
        Nitrogen,   33.92
                                ———
                              100.00

Martin[*] has devised a method of ascertaining the percentage of nitric acid, by its conversion into ammonia. Nitric acid when mixed with sulphuric or muriatic acids, in the presence of metallic zinc, is converted into ammonia (Gmelin I, 828). By placing some zinc in a mixture of the two acids, there is no disengagement of gas, whilst the nitric acid is converted into ammonia. Hydrogen in its nascent state combines with the oxygen of the nitrogen compound, produced by the nitric acid alone.

Metallic zinc, with dilute nitric acid, gives protoxide of nitrogen; and by taking one equivalent of this gas and four equivalents of hydrogen, water and ammonia may be formed.

$$NO + 4H = NH^3 + HO.$$

The nitric acid, acting gradually and slowly on the zinc, is transformed into ammonia, equivalent for equivalent. When this reaction has ceased, then follows a disengagement of hydrogen gas from the zinc, which is permitted for a few seconds. It now remains to ascertain the percentage of ammonia. The ammonia may be distilled off and then absorbed by a normal or previously ascertained quantitative solution of oxalic acid, and afterwards to ascertain the quantity of oxalic acid not taken up; deduct this from the original quantity contained in the absorbing solution, and the result gives the percentage of oxalic acid neutralized by the absorption of the ammonia; from this the ammonia is calculated. Mohr's apparatus for the disengagement of ammonia may be used with advantage in this operation. See Mohr's Traite d'analyse chimique, supplement, p. 402, Paris, 1857.

Tilberg[†] analysed the Stockholm Nitro-Glycerin with the following results: $C^3 H^5 (NO^2) O^3$
(the Carbon atoms being estimated as 12, Hydrogen 1, Oxygen 16,) and regarded it as Mono-Nitro-Glycerin.

In proof of the fact of Nitro-Glycerin being explosive by concussion effected at a distance, if proof were needed, I instance a small can containing about 4 lbs. of Nitro-Glycerin left by the blaster about 350 feet from the heading, and partially protected by the rail which was curved upwards to prevent the cars running over the dump, was exploded, when a full charge of 16 holes was fired in the heading at the West End of the Hoosac

---

[*] Comptes rendus, V. xxxvii, p. 947.
[†] Chemical News, March 1869, p. 151.

Tunnel. It will be noted that there could be no heat developed 350 feet from the primary explosion, and being enclosed in an ordinary kerosene can, it appears a striking instance of the possibility of explosion from induced concussion.

Again, in April, 1872, a cartridge of Nitro-Glycerin was left in the cartridge chest, containing about 2 lbs. Nitro-Glycerin, whilst 20 charges of blasting powder were fired in the heading, 200 feet distant; the explosion of the powder was unusually heavy, and the Nitro-Glycerin exploded, tearing the chest to pieces, fracturing the air main and disrupting the track. This indubitably proves the explosion of Nitro-Glycerin by concussion, and should warn every operator to be careful to place any surplus explosive away from exploders, and as far distant as possible from where an explosion is intended, and particularly in such position that if it should explode, a contingency possible, there may be no one near the vessel containing such surplus.

*The experiments of February 17, 1870, described by Professors Barker and S. W. Johnson, where water and glass intervened to receive the heat and concussion, confirm the fact of Nitro-Glycerin being explosive by concussion, without heat or pressure; in these instances neither heat nor pressure were admitted, yet the explosion blew the tub into fragments, cutting off the staves level with the hoops, smashing and fracturing the bottom of the tub on the rock serving as a pedestal, and sending the water up so that it descended in a shower seventy feet from the point of explosion.

It is proper I should here announce that, after a series of experiments, during my leisure hours, extending over several years, with nitro-mannite, nitro-sugar, nitro-dextrin, nitro-starch, and nitro-naphthalin, with a view to obtain a homogeneous compound convertible wholly into gaseous matter, and miscible with liquid Nitro-Glycerin, which would not explode under ordinary conditions, I have succeeded in obtaining such a mixture, viz.:

Nitro-Glycerin, thirty parts.
Nitro-Toluol, ten parts.

Mixed, this will not explode when struck on an anvil, burns when thrown on to the fire, and can only be exploded with very heavily charged exploders, containing, say, fifteen grains of fulminate, better and more surely, however, with twenty grains. To this I know but one drawback: it does not solidify at a moderate

---

* See abstract of Prof. Barker's affidavit, towards the close of this pamphlet.

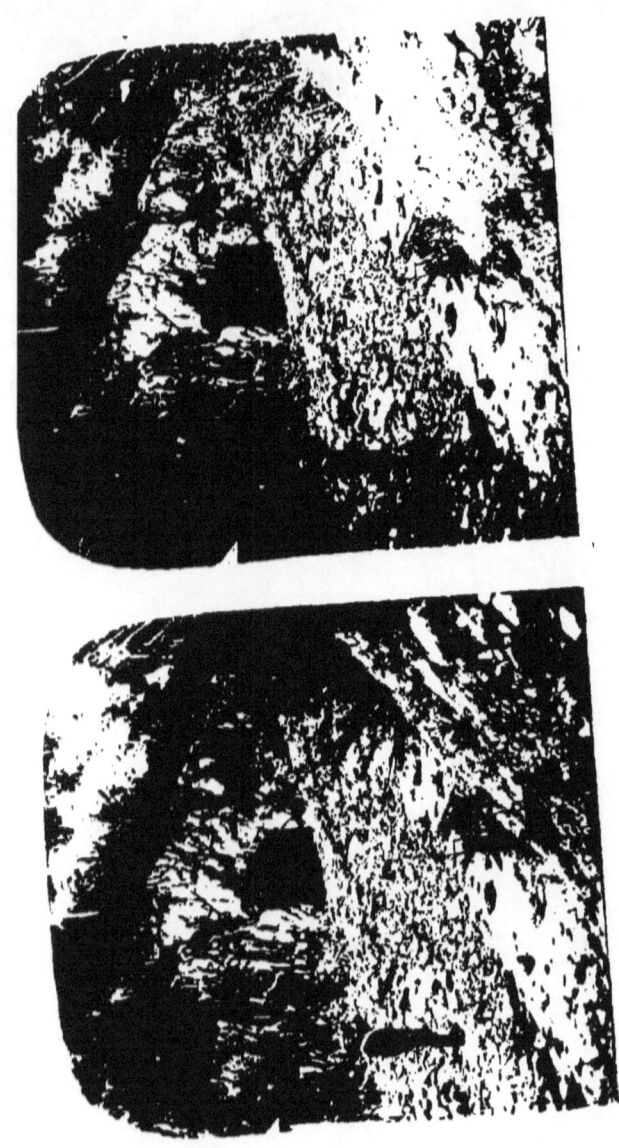

(45° F.) temperature, and, if the containing vessel should leak, a too frequent source of accident with inferior Nitro-Glycerin that cannot be congealed, the nitro-toluol is liable to evaporate, and the Nitro-Glycerin is then left with its usually dangerous properties unimpaired.

This was patented by C. Volney, who formerly blasted for me, and for the Lake Shore N. G. Co., and assigned to me for a consideration.

CHAPTER IV.

Electricity for Blasting Operations.

Although half a century has passed since blasting by electricity was effected by Col. Pasley, in his submarine explosions for removing the wreck of the Royal George, at Spithead, the apparatus for exciting the electricity necessary to explode many charges simultaneously, is still (May, 1872), very unsatisfactory. Mr. H. Julius Smith, of Boston, taking the Austrian friction battery, recommended by Baron Abner, in his report at Vienna, for his basis, has ameliorated the arrangements by enclosing the working parts in a better vulcanite casing, and securing the discharge by reversing the motion of the handle, but the objections remain that an ebonite plate is scratched by the rubbers, that specks of the sulphuret of tin, used as an amalgam, cause a partial discharge all over the surface of the plate, rendering it a short-lived machine whose power is limited, unless the priming of the exploders is made very sensitive, and liable to explode by atmospheric electricity. Several fatal accidents have occurred to miners, from premature explosions of the charge whilst loading the holes, and these fatalities having been traced to the "over-sensitive priming" used, it behooves the mining engineer to look well to the exploders offered him, and in every case he will find where cotton immersed in a varnish is sufficient insulation to protect the wire from losing its electricity, the priming used for charging such exploders is too dangerous for miners' use, and involves a grave responsibility.

Mr. Abel's Electro-magnetic Exploder limits the discharge to a series of five mines, or blasts in each series, being the Verdu or Savare system, and involves several leading wires for numerous explosions, and although yielding electricity in quantity it lacks intensity.

The Holtz machine is altogether too vicarious in its operation for blasting purposes. A machine or apparatus that will discharge 100 blasts, if needed, durable, and not liable to derangement or wear, is a necessity, and it should evolve enough electricity and of sufficient tension to jump between the wires 1-20th of an inch apart, necessary to fire priming, so as to secure simultaneous firing. The heated wire, or a quantity of electricity heating wire by the resistance a small wire offers to the current, since it occupies time, brief though it be, involves, as I think, the objection that the discharges cannot be simultaneous in, say twenty blasts. Of this class are the machines now in course of construction by Mr. Moses Farmer, of Boston, where the exciting power is manual labor, being a dynamo-electric machine. Breguet's electro-magnetic exploder, giving a spark by breaking contact, is altogether too weak, at least for the American contractor.

The ordinary Ruhmkorff coil is accompanied with the objection, that in a numerous series of blasts, the spark, when it has passed some five or six holes, seems to vanish in a glow, and to lose the heat necessary to effect decomposition of the priming, besides the incumbrance of acids and battery; in brief, it is not sufficiently portable for the use of contractors.

During the past four years I have given this subject much attention, and, having experimented pretty extensively, I have secured the first point, viz.: a safe priming which is not affected by the induced electricity caused by machinery running, friction of handling, or atmospheric electricity. My present aim — the evolution of electricity of sufficient intensity to leap fifty to one hundred solutions of continuity, i. e., effect fifty blasts simultaneously, I hope I have secured, but this subtile force, electricity, is so readily affected by so many interfering elements in blasting operations, that it would be premature in this patent-demanding age, to communicate the progress I have obtained, until the several apparatus I am now constructing (three forms of machine), are complete, and have been subjected to actual work in severely critical hands. An inventor is no judge of the success of his own bantlings.

Aware of the short life of the frictional electric machine, as at present constructed; knowing how the ordinary induction coil diminishes its intensity of spark, in proportion to the number of blasts to be fired; seeing that the Electro-magnetic machine is limited to a series of five blasts, which can only be exploded consecutively; that the Electro-dynamic machines are open to this last objection, besides destruction of their conducting parts by over-heating, whilst in the matter of adopting "over-sensitive priming" to compensate for the deficiency of electricity or cheap conducting wire, there looms up the danger to the miner of handling exploders, which "go off by looking at" them, it seemed that, unless some amelioration was effected in these details, the great economy of simultaneous blasting by means of electricity would have to be abandoned. Add to these difficulties the fact that any casualty occurring from any of the above causes would reach the public as caused by Nitro-Glycerin, and my reader will comprehend the interest I have felt, during the past four years, in solving the following problem:

To construct an apparatus that will, under every condition of atmosphere, whether damp, dense or rarefied, evolve, at the will of the operator, abundance of electricity; such electricity to possess the property of developing intense heat, so as not to need a very sensitive priming, and to possess sufficient tension to overleap numerous solutions of continuity, say fifty, at a flash. Next, to discover a priming composition, to insert between the solutions of continuity, that would not be affected by moisture, that would bear handling without danger of exploding, be unchangeable for years, unaffected by the induced electricity of the atmosphere, whether caused by thunder storms, lightning on the rail, machinery belting in motion, or steam blowing off from a safety valve, ozone, etc., and yet not too exhaustive of the electric force of the spark required to fire it.

The above seemed to me the conditions necessary for the apparatus and the exploder in firing with electricity.

In addition to these, for conducting such electricity to the points required, the best conductor, and the best insulation attainable.

Further, that as Nitro-Glycerin was an expensive explosive to waste, to supplement the above details with some material that would absolutely develope its extreme force instantaneously, and not as is easily the case, burn a part, explode a part, and throw the remainder into the atmosphere, to poison the miners, or by missing fire, endanger life, and waste time. How these objects, so desirable, have been obtained, I now proceed to relate.

By modifying the ordinary induction coil, so as to make it yield a highly heating spark, and remedying its property of losing tension rapidly after leaping four or five solutions of continuity, the Messrs. Ritchie & Sons, of Boston, have constructed for me a coil that fires 18 intervals when charged with rifle powder simply; and they are now constructing another coil capable of firing fifty mines, when charged with priming that is perfectly safe to handle, and fulfilling the conditions enumerated above. One spark alone is required to effect these results, which may be summed up as "eliminating the heating properties of induced electricity."

I have previously referred to the necessity of using a heavy charge of fulminate of mercury, in order to secure perfect and instantaneous explosion of a charge of Nitro-Glycerin, without confining the latter; the manipulating this explosive salt (fulminate of mercury) without hazard to the operators (generally girls), was accomplished by precipitating gum mastich from its alcoholic solution, by the addition of water, and mixing in the moist fulminate, and then filling the pasty compound into a stout copper capsule, which is subsequently enclosed in a wooden case, saturated with paraffine. The resistance of the stout copper capsule, immensely adds to the effective force of the exploder, and ensures the most effective explosion of the Nitro-Glycerin, which cannot be obtained by a wooden capsule alone. These details as to the requirements for effectively exploding the nitro-compounds, have been very fully examined and proved, by Abel, Article, Pyroxylin, Watts' Chem. Dictionary, Vol. 4, p. 776, et seq., and daily use confirms them. My observation of the fatalities that have occurred with over-sensitive priming composition, introduced with a view to compensate for deficient electric force, and thus to permit the use of a weak battery and cheap cotton

covered wire varnished over (instead of gutta-percha insulation), in order to substitute a weak current that would be sufficient to fire these over-sensitive exploders for the stronger current required to fire a safe priming, satisfy me that electric blasting had better be discontinued, and tape fuse resumed, unless the work will bear the expense of absolutely safe materials. Better to face the difficulty, construct efficient electric apparatus, convey the electricity along wires of perfect insulation to a safe priming, and by complete and violent explosion of the Nitro-Glycerin, or powder, make such effective blasting as not to throw away the labor of drilling, candles, power, and blasting materials. I believe this the true economy. These details may seem wearisome, but the casualties of blasting can best be diminished by avoiding missed holes, a result only attainable by using materials absolutely reliable; and the reader, if he has ever attempted to harness up as a team those subtile, evasive, terrific forces—electricity and explosives, for the service of his fellowman, will excuse the writer's earnestness and agree with him that in such a task the rule should be "Aut nunquam tenta aut perfice."

CHAPTER V.

The Tri-Nitro-Glycerin Manufactured at the Hoosac Tunnel — How Tri-Nitro-Glycerin is Made — How Stored — How Gutta-Percha is Purified — How the Conducting Wires are Insulated — How the Exploders are Manufactured.

There are probably few of my readers who have ventured to trust themselves within a Nitro-Glycerin manufactory; the very name is sufficient to make the passer-by quicken his step, till he is a safe distance beyond the dreaded precinct. Some account of such a factory will, accordingly, be interesting to many who are familiar with the article, perhaps have used it, but whose curiosity has not been of such a nature as to induce them to pay a visit to the works, where the least negligence involves a death penalty.

About 100 yards beyond the West shaft of the Hoosac Tunnel, is to be seen a board fence surrounding about ten acres of ground, with the announcement, "NITRO-GLYCERIN WORKS;— DANGEROUS;— NO VISITORS ADMITTED."

A drive leading between two rows of buildings brings the "visitor" to the acid house, a well-ventilated building, 150 feet long. Here are 11 stills, each seven feet long and two feet in diameter. Under these a light, slow fire burns, which is carefully attended to, for the temperature must be kept moderate. In each of these stills is placed 300 lbs. of nitrate of soda and 375 lbs. of sulphuric acid. A stoneware pipe conducts the gases, at a temperature of about 180° F., from each still into a stone receiver or condenser, or rather a series of four condensers connected by stoneware pipes, ranged on a platform three feet above the ground. Into the first of these 150 lbs. of sulphuric acid is poured, into the second 150 lbs., into the third 100 lbs., and the fourth is empty. The nitrous vapor passes from the still to the first condenser, where a portion of it, forming as it condenses nitric acid, is taken up by the sulphuric acid; the remainder passes on to the second, third and fourth condensers, though a very small portion is left to pass into the last, which only requires to be emptied once a month. It takes about twenty-four hours for the still to complete the conversion of its contents into nitric acid, at the end of which time the resultant mixture of acids, about 600 lbs., is run off into carboys, twelve of these being filled from three stills. About 100 carboys are generally kept in stock, as the acid does not spoil when kept closed. These carboys are then emptied into a soapstone tank having a capacity of 18 carboys, and an iron pipe, connected with the main leading from two blowers in the engine house, is inserted into the acid, causing a current of air to agitate it so as to remove the nitrous fumes, mix it thoroughly and bring it all to uniform strength. Formerly, this was effected by removing the acid into a glass vessel containing about forty gallons, and it required boiling for hours; the mode now practiced occupies only five minutes and the risk of fracture of a glass vessel in a sand bath is avoided. The acid is then carried into the converting room, about one hundred feet long and well lighted, where it is weighed, seventeen pounds being poured into each of one hundred and sixteen stone pitchers which are arranged in nine wooden troughs placed in the centre and at the end of the room, and these troughs are now filled with ice-cold water, or ice and salt, so as to rise within four inches of the top of the jar. On shelves above the troughs, are arranged glass jars, one to each stone pitcher. Into each of these glass jars, two pounds, by weight, of pure Glycerin is poured, and this, by means of a siphon, with a rubber tube attached, about two feet long, falls drop by drop into the corresponding pitcher of mixed sulphuric and nitric acids. Immediately below the shelf, in which the Glycerin jar stands, is a $2\frac{1}{4}$ inch iron pipe, which brings a current of cold air from the receivers connected with the two blowers before mentioned. This current of air is distributed to each jar, while the acid and glycerin are mixing, by a rubber pipe, to which is attached a glass tube 16 inches long, and with a $\frac{1}{4}$ inch bore. During the hour and a half to two hours that the glycerin takes to run off into the pitchers, the greatest care, and the closest attention is requisite. The three men whose duty it is to attend to the mixing process, have each a row of pitchers to watch, walking the whole time up and down, beside them, with thermometer in hand, and as the nitrous fumes rise from the forming Nitro-Glycerin, they stir the mixture, with the glass tube before mentioned, in any pitcher that may be giving out too violent fumes. Sometimes this is caused by the glycerin running a little freely, which fires the mixture, wastes the glycerin, forming oxalic acid, and developes unpleasant vapors. In such a case, by pushing back a little wooden peg in the glass jar, the flow of glycerin is lessened, and by stirring with the glass tube the nitrous vapors dispelled. Should the engine also stop working by any unforseen circumstance, the current of air will of course be stopped, when the mixture will take fire. In this case, it is necessary to stir the mixture, and at once stop the flow of glycerin. When the glycerin and acid is all mixed, and the nitrous fumes cease to appear, the Nitro-Glycerin from each pitcher is dumped into a large tank of water, at a temperature of 70°, about 450 lbs. of Nitro-Glycerin being the amount of each batch manufactured. The Nitro-Glycerin sinks to the bottom and is covered by about six feet of water. Here it remains for fifteen minutes to be subsequently washed free from any impurities. This tank goes through the floor into a basement chamber, its bottom being on a slight incline, so that the Nitro-Glycerin may run out easily. The water is first drawn off from the top of the Nitro-Glycerin, and then the latter is run into a wooden swinging tub, in shape somewhat like an old-fashioned butter churn, but a good deal larger in diameter. In this it is washed five times, three times with plain water, and twice with soda, a current of air working through it at the same time. The water from this tub is run off into a wooden trough, which conveys it to a barrel buried in the earth, in the side of which a hole carries it to another barrel a little lower down the hill, and this again to another barrel, whence it finds its way to the dump of rocks being removed from the tunnel, any Nitro-Glycerin that may have escaped in the washing process being collected and retained in one or other of these barrels.

The Nitro-Glycerin is by this time thoroughly washed and ready to store in the magazine, 300 feet distant, to which it is carried in a couple of copper pails at a time, by a man with a yoke, similar to what milkmen use for carrying their pails. Curious thought, that a man carrying a couple of harmless looking pails with only a little colorless fluid in them, should have enough explosive matter about him to annihilate a regiment.

In the magazine the Nitro-Glycerin is poured into "crocks," as they are called, earthenware jars holding 60 lbs. These crocks are then placed in a wooden tank $2\frac{1}{2}$ feet deep, which holds 20 of them, and immersed to within six inches from the top of the jars in water warmed by a small pipe from the boiler, to raise the temperature to 70°, at which temperature it is kept all the time, as nearly as possible. They remain in this water for about 72 hours, during which time any impurities still remaining rise to the surface as scum, and are skimmed off with a spoon. The Nitro-Glycerin is then chemically pure, transparent as water, refracts light powerfully, and is ready for packing. The tin cans, lined with paraffine and containing 56 lbs. each, are placed in a shallow wooden trough, and the Nitro-Glycerin being poured from the crocks into copper cans, is again poured into the tins through a gutta-percha funnel, the bottom of the trough being covered with a thick layer of plaster of paris,

which absorbs and renders harmless any drops of Nitro-Glycerin that may be spilt. The tins when filled are then placed in a wooden trough containing iced water, or ice and salt, where the Nitro-Glycerin is slowly crystallized or congealed; in this condition, it is stored away in small magazines 300 feet distant, in amounts of 30 to 40 cans each, until required for use.

When the Nitro-Glycerin is to be conveyed over the mountains, the tins are packed in open wooden boxes, with two inches of sponge at the bottom, and four rubber tubes underneath; these are long enough to allow the ends to come one inch over the top of the tin on opposite sides, thus interposing two elastic tubes between the outside of the tin and the inside of the wooden box, rendering it perfectly safe to carry. Each tin is cellular, i. e., from the top of each tin to the bottom a tube passes, about ten inches deep and $1\frac{1}{4}$ inch in diameter, for the purpose of thawing the congealed Nitro-Glycerin when the blaster is ready to use it, liquefaction beeing effected with water of 70° to 90°. The tins being closed with a cork wrapped in bladder, are put into a sleigh or wagon, covered in summer with a layer of ice and blankets, and may thus be carried any distance in this purified crystalline state, as safely as so many tubs of butter.

The reflecting reader will note the care taken to purify the Nitro-Glycerin; it occupies $1\frac{1}{2}$ hours to make it, about 72 hours to purify, and about 48 hours to congeal or crystallize it. And yet there are parties who attempt to make and vend Nitro-Glycerin, and induce miners and contractors to use it, taken direct from the precipitating tank, with all its impurities tending to decomposition, and requiring only time and moderate temperature for spontaneous explosion; hence, I believe many accidents.

Proceeding back to the factory, two ice-houses will be noticed, capable of containing 400 tons of ice, required for crystallizing Nitro-Glycerin in summer. There is a small engine-house with a boiler of fifteen horse power, and engine of about ten horse power; this latter, to pump water into the washing tank, run the two "blowers," and give power in the gutta-percha factory. The air is not pumped directly into the pipe which distributes it to the pitchers, as the pressure would not be always uniform; but into two receivers under the floor of the factory, whence it is evenly distributed, and deprived of watery vapor, which if blown into the pitchers would raise the temperature and vitiate the product.

Attached to the factory is a building about 90 feet long, for covering the copper wire (used in exploding) with gutta percha, so as to render the insulation perfect. The first process is to purify the crude gutta-percha which is imported in blocks about a foot long. This is placed against a rasping machine with toothed knives about four inches apart, which crush and tear the gutta-percha to pieces, delivering it into a trough of water. The impurities sink, while the gutta-percha floats. It is then warmed in a steam jacketed kettle, and when still plastic is put into another tearing or rasping machine with another series of knives set closer together, from this it drops into a trough of clean water, more dirt separating. This is repeated two or three times, as it is most important that no extraneous matter should be retained in the gutta-percha, because it would interfere with perfect insulation, and so place in jeopardy the lives of several men. It is again steamed and put into a "masticator" consisting of a fluted roller working in a steam jacket; here it is "chawed up" for about six hours, until it arrives at a proper consistence; it is then passed between two smooth cylinders heated by steam, and transferred thence into a cylinder, where it is pressed through gauze wire, under a pressure of four tons to the inch. Being thoroughly cleansed, it is then steamed, masticated and pressed between the cylinders, and is ready to cover the copper wire. Five wires at a time, horizontally parallel to one another, are passed through a gun metal mould with a disc at the further end perforated with five holes but little larger than the wires themselves, placed at the base of an upright cylinder. The gutta-percha is inserted in the top of this cylinder, and a pressure of 95 tons is put upon it by means of a screw, when it is pressed into slots in the mould surrounding the wires, which are then drawn from the holes in the disc, through a trough of water 80 feet long, and back again: it is then wound on drums ready for use. The "leading" wire receives two coatings, separate discs having larger bores being attached to the brass cylinder.

A house is attached to the factory, for the foreman and his family.

Perfect system pervades this factory, and is absolutely necessary in the manufacture of Nitro-Glycerin, to ensure safety. The steadiest men possible are selected for the work, and the foreman of the gutta percha department, Mr. Robert Wallace, who has charge of the machinery, is a skilful machinist and a thoroughly trustworthy Scotchman. He has four sons employed, of whom one takes charge of the works at Maysville, Kentucky, another, is foreman of the Nitro-Glycerin factory.

Three men are employed in the acid house, working in three shifts of eight hours each, but they do not actually work more than seven hours; every movement is like clock work, every man has his place and special duty, which he is expected to perform at the proper time. In the morning, at 7 or $7\frac{1}{2}$ A. M., two men dump the carboys of acid into the soapstone tank and mix them, while a third is filling the glass jars with glycerin. This operation takes about an hour. One draws the acid, another weighs it, and a third carries it to the troughs. After an interval during which the acids cool, three men attend closely to the converting of glycerin into Tri-Nitro-Glycerin, knowing that their safety, and the safety of every man on the works, depends on themselves alone, during this process. After the Nitro-Glycerin is dumped into the water tank, two men are employed in washing it, down stairs, while two wash the stone pitchers with water; more water, temperature about 60°, is swilled on the floors so as to keep them scrupulously clean and perfectly free from atoms of Nitro-Glycerin, which, stepped upon while the men are at work, might send them to eternity, and the building to smithereens. The room is then prepared for next day's operations, and by about one or two o'clock, after six, or at most seven hours' work, the day's task is done. Mr. Wilson, in charge of the purifying process, canning, and preparing for shipment, has now been over four years at this work.

Making exploders is a distinct operation, requiring great precision. The materials of which the priming for fuses is composed, are prepared in my private laboratory, and consist of sulphide and phosphide of copper with chlorate of potash. Considerable nicety of manipulation is required to prepare the former of these compounds so as to obtain homogeneous, uniform sulphides and phosphides, and, from the failure of several chemists — and some of our best have attempted the manufacture — to prepare them, I attach great importance to this work, invariably making them myself. For, if prepared with the above ingredients no accident can occur from atmospheric electricity, friction etc., a contingency which all other primings now in use are liable to. The priming is then taken to the warehouse where from three to four hands are employed in making it up into exploders. Two insulated wires from 4 to 12 feet long, are inserted in the smallest end of a wooden tube, previously dipped in boiled paraffine, $\frac{3}{4}$ inch long and $\frac{1}{8}$ inch diameter at one end, and $\frac{3}{16}$ at the other, to which they are fastened by a shoulder of gutta-percha. Immediately before the priming is inserted, an electric spark is passed through and between the wires where the priming is put so as to ascertain that the insulation is perfect, and to guard against the possibility of a miss-fire. This being proved, the priming is put in at the other end of the tube, and a small paper plug boiled in paraffine inserted; then a copper cap, $\frac{3}{4}$ inch long and $\frac{3}{8}$ inch diameter, receives 20 grains of fulminate of mercury, on the top of which a varnish is poured which prevents any of the fulminate from being shaken out by accident, or affected by vibration. This copper cap is then placed in a larger wooden cap $1\frac{1}{2}$ inch long, the fuse inserted about $\frac{1}{4}$ inch, when it fits tight, the wooden part painted with asphaltum varnish around the joints, and the exploder is complete and ready for service. Three hands employed ought to make 1,000 a day of these exploders.

Having thus given a full account of the manufacture of Nitro-Glycerin and its appurtenances, I will conclude with the remark that there is no danger in the manufacture when due precaution is used; but, to paraphrase the language of Professor Tyndall, in his preface to "Hours of Exercise in the Alps": "For rashness, ignorance, or carelessness, Nitro-Glycerin leaves no margin; and to rashness, ignorance, or carelessness, three-fourths of the catastrophes which shock us are to be traced."

CHAPTER VI.

Explosive Mixtures.

The laws of nature are immutable. To-day, to-morrow, forever—unchanged, unchangeable, as the great Creator himself, who established them, and it is only from scientific research, starting with the conviction that these laws are God's laws, and therefore immutable, that results of general utility can be obtained. Believing that everything which, in common parlance, is termed "an accident," is simply a violation of these laws through carelessness or ignorance, it is the duty of the scientific chemist to investigate the causes and effects of the adherence to or violation of these laws in regard to the science of which he is a student. As a chemist I have accordingly applied myself to a close examination of the phenomena attending the preparation and use of Nitro-Glycerin, and consequently to the investigation of the mixtures purporting to be substitutes for Nitro-Glycerin and gunpowder, of which Nitro-Glycerin is the active base.

And this brings before me, in all their glaring defects, the anomalies of the patent system of our country, especially in regard to chemical compounds. For the past hundred years, the greatest chemists the world has ever known, have given the results of their researches free, and untrammelled by any patents, though they might, indeed, have justly taken toll of the world at large for their discoveries. I need only instance Berzelius, who threw open to the world the numerous discoveries of his long and valuable life, and Pelouze, the celebrated French chemist, who devoted fifteen years of his life to the investigation of the constituents of fatty matters and their decomposition into stearic, margaric, oleic acids and glycerin. Let the reader picture to himself, for a moment, what would have been the state of affairs in the manufacturing world, had all the chemists of the last fifty years patented every discovery they made, every mode of preparation they suggested; how dark, gloomy and uncertain would the path of our manufactures have been; they must almost have stood still until these patents, and perhaps their renewals also, had expired. By such a course, the bleaching and printing of cottons, and all the numerous processes dependent on applied chemistry, would have been deferred half a century; for it is only by the quick, free application of the discoveries of the unselfish chemist, that the progress that has been made was possible. What a contrast to the self-aggrandizement of the present race of patent-seeking chemists! An individual, with the labors of the grand army of scientific chemists for the past hundred years before him, selects one, two or three chemical compounds, mixes them, modifies to a certain extent some property of either of them, applies for, and obtains, a patent. Then for seventeen years this "ghoul" sits over his mixture, and, with the assistance of a lawyer, proceeds to blackmail any one, who, in attaining certain results, is led by the properties of the several compounds to avail himself of a similar mixture. The discovery of a Sobrero is attempted to be appropriated by a Nobel and his assignees, and, with the confidence inspired by the weakness of a patent examiner, who chuckles at the delusion of the patentee, they absolutely infer that, because they have a patent, they can appropriate the result of the chemist's labors obtained 20 years before. The patent office secures $35.00, the examiner his salary, and the ceilings of the noble building at Washington are ultra-marined, until the visitor's eyes are dazzled with the brilliant color. Finally comes a suit in chancery, in which thousands of dollars are expended, and in which these stealers of other mens' brains, count less on their claim than on the hope that they may so interfere with their opponent's occupation, and so deplete his pocket with law-costs, that he will submit to accept a free license, at least, and thus enable them to terrify others into payment.

The above remarks are somewhat of a digression from the subject of this chapter, but, I think most of my readers will admit that they are by no means uncalled for. I have been told, and the newspapers teem with assertions, that these patented explosive compounds, with high sounding names, will bear "tamping" as hard as gunpowder, are safer, more powerful and cheaper than Nitro-Glycerin. We are a people, Barnum says, who like to be humbugged; I am afraid we are not the only people who like to be humbugged — it is a weakness of humanity — but this I do believe; the man who is addicted to humbug, had better give Nitro-Glycerin a wide berth, that is, if he hopes to end his days on a feather bed.

Let us briefly examine these patents — the Lord deliver us from all such — for explosive mixtures, and see the amount of invention required.

For a mixture of Nitro-Glycerin with rotten-stone, a patent was granted, and (the name being the only real invention) it was called "dynamite." *

Make a mixture of Nitro-Glycerin and sponge, and patent it, and forthwith " Porifera nitroleum " is presented to an admiring public. †

Add plaster of Paris to Nitro-Glycerin, patent it, and you have in all its explosive power, "Selenitic Powder." ‡

Try red lead and Nitro-Glycerin together, and when patented, " Metalline Nitroleum " is the last new sensation to astonish the weak nerves of contractors. §

Take some gunpowder in a fine state of division, and moisten it with Nitro-Glycerin until it becomes " the color of mud and

---

* "Dynamite" – Patent No. 78,317, dated May 26, 1868, granted to Alfred Nobel, of Hamburg, Germany, assignor to Julius Bandmann, of San Francisco, California. The following is the substance of the claim: " My invention consists in combining with Nitro-Glycerin a substance which possesses a very great absorbent capacity, and which at the same time, is free from any quality which will decompose, destroy, or injure the Nitro-Glycerin, or its explosiveness. The substance which most fully meets the requirements above mentioned, so far as I know, is a certain kind of silicious earth, known under the various names of silicious marl, tripoli, rotten-stone, etc."

† " Porifera Nitroleum " – Patent No. 93,753, dated Aug. 17, 1869, granted to Taliaferro P. Shaffner, of Louisville, Kentucky. The claim is as follows: " I claim a compound composed of a mixture of Nitro-Glycerin with sponge or other vegetable fibre."

‡ " Selenitic Powder " – Patent No. 93,752, dated Aug. 17, 1869, granted to Taliaferro P. Shaffner, of Louisville, Kentucky. The claim is as follows: " I claim the combining of nitroleum or Nitro-Glycerin with plaster of Paris, or equivalent substances, in such manner as will make an explosive compound."

§ " Metalline Nitroleum " — Patent No. 93,754, dated Aug. 17, 1869, granted to Taliaferro P. Shaffner, of Louisville, Kentucky. Claim as follows: " I claim a compound composed of a mixture of Nitro-Glycerin with metallic powder or atoms, however formed or produced."

about the consistency of putty"; assure the editor of the Barnumtown Inquirer, that it has five times the explosive power of Nitro-Glycerin, and forthwith a flaming article appears, upon the new explosive agent, "Lithofracteur." *

Make a compound of sawdust and Nitro-Glycerin, and let your patent prove that you are unacquainted with the commonest properties of sulphuric acid and charcoal, that, on the face of it, your preparation cannot possibly be made as you describe (that is not the business of the examiner, or if it be, he is so bothered by Prussian officers that these facts escape his notice), on payment of $35.00, a patent will issue, give it a name, say, "Dualin", boldly assert that its properties are unequalled; let a governor of a state, whose experience is confined to fire-crackers, witness an explosion (it is not material what substance you explode before him), hire a steamer, give a splendid collation, invite all the reporters within reach, make any statements you please to them (they will be swallowed along with the collation, especially if washed down with plenty of Heidsick), and there is no telling where this halo of a patent may not carry the unscrupulous patentee. †

But these assertions involve loss of life, as, for instance, when Joseph Butloe was killed at the Hoosac Tunnel. He was attempting to introduce a dualin cartridge into a drill-hole, and as it did not reach the bottom of the hole he endeavored to push it in further with a "tamping stick," a method which the inventor of dualin advocated, and regarded as perfectly safe. Unfortunately, however, in the present case it was not so, the explosion following the first "tamp" instantly killing the operator, and exploding the mis-statements of the patentee.

Truly, these gentlemen are wonderful mathematicians; they have discovered that a part is greater than the whole, that various mixtures of inert matter with Nitro-Glycerin, have greater explosive power than Nitro-Glycerin per se.

As Dualin is the only one of these compounds that has been attempted to be brought in any way into competition with Nitro-Glycerin, in the Eastern States, a synopsis of the results may

---

* "Lithofracteur"— For a wonder this has not been patented.

† "Dualin"— Patent No. 98,854, dated January 18, 1870, granted to Carl Dittmar, of Charlottenberg, Prussia. Claim as follows: "I claim a compound consisting of cellulose, nitro-cellulose, nitro-starch, nitro-mannite and Nitro-Glycerin, mixed in different combinations, depending on the degree of strength which it is desired the powder should possess in adapting its use to various purposes."

possess interest. Some six different parcels of dualin in all, have been experimented with at the Hoosac Tunnel, and of these the first shipment, being useless at the West End, was forwarded to the Central Shaft, and there again tried, but the effects, as compared with the Nitro-Glycerin supplied by the writer, were not such as to justify the contractors in continuing its use, consequently it was thrown out. Another parcel, intended to be stronger, shipped in the hot summer of 1870, exploded in the cars in transit at Worcester, proving, what had been suspected from a perusal of the dualin patents, that the inventor was really ignorant of the properties of the materials of which his combination was composed. From evidence adduced at Worcester, given by the compounder of dualin, and also by a manufacturer of exploders, some of whose wares were in the same car, it appeared that the Nitro-Glycerin exuding from the mixture of sawdust (40 per cent.) and Nitro-Glycerin (60 per cent.) of which the dualin, made at that time by Mr. Dittmar, was composed, flowed in a pool on the floor of the car, and, when the cars were set in motion, a series of sharp detonations ensued, probably from this pool of Nitro-Glycerin running on to the wheels and being compressed or hammered during the revolution of the car wheels on the rails, firing the pool, which in turn fired the whole shipment of dualin, together with the exploders.

After some months further shipments were made, and in all cases the trials made with these were superintended by the introducer of dualin, and, in every case but one, were reported failures, and rejected. In the case in which a success was reported, a small parcel only was brought along, and exploded side by side with Nitro-Glycerin; that is, four holes were charged with dualin, and four other holes nearly parallel with them were charged with Nitro-Glycerin. The enlargement was brought down, but whether the work was principally done with Nitro-Glycerin, and only partially by the dualin, was left to conjecture. The foreman of the drillers asserted that the side charged with dualin was seamy, whilst the side containing the Nitro-Glycerin was solid, and without any seam. However, it was claimed by the inventor that dualin was now a success, and a further trial, viz.: the sixth, was undertaken, and 1,500 lbs. of dualin brought on the ground, about the 26th of November, 1870. On Tuesday, the 28th, the experiments under the supervision of

Mr. Dittmar commenced, and were continued on the 29th and 30th, but they demonstrated beyond cavil, there being no Nitro-Glycerin fired at the same time to assist them, that dualin was of "no account," not one single hole having been "bottomed," and, again, the dualin left over from this experiment, 1,300 lbs., was thrown out, as utterly unable to effect the blasting results obtained by the Nitro-Glycerin it was brought to supersede. Four hundred pounds of this was ordered to the Central shaft, but the results at the East End being so conclusive, it was consigned, like all the previous shipments, to the tomb of the Capulets, and was subsequently used up for trimming, in lieu of powder.

In a previous chapter, I gave a full account of the experiments made at Hallett's Point, New York. On that occasion, General Newton, of the United States Engineers, reported to me that he considered that Nitro-Glycerin, in point of economy and power, had the advantage over both dualin and powder even when supplemented by fulminating fuse. The advantages claimed (only by the inventor) for dualin, are, that it is cheaper, safer, and more powerful than Nitro-Glycerin, and some experiments made in Prussia, are adduced in proof. I have to observe, on this point, that the Nitro-Glycerin made by the Nobel process, probably used in Prussia, is very inferior to the Tri-Nitro-Glycerin made by my process, both in stability and in explosive force, and it is much more readily exploded, fifteen grains of fulminate of mercury being necessary to ensure explosion of this latter, without chance of failure. Nobel's Nitro-Glycerin is said to expand when solid, in which state the slightest friction is said to explode it, while Mowbray's Tri-Nitro-Glycerin actually contracts about one-tenth in bulk when solidifying, and cannot be exploded when in the solid state, except by a heavy charge of fluid Nitro-Glycerin fired with it. Nobel's preparation is yellow, and gives off nitrous fumes, and is claimed by the patentee to solidify at 50° F., while Mowbray's is colorless as water and solidifies at 45° F.

It may be possible, but not probable, therefore, that Nobel's Nitro-Glycerin is inferior to Dittmar's dualin, as used in Prussia; the latter then said to have been a preparation of nitrate of ammonia, sawdust immersed in sulpho-nitric acid and Nitro-Glycerin: but that 40 per cent. of washed sawdust (not treated with sulpho-nitric acid), moistened with 60 per cent. of a dark colored and evidently impure Nitro-Glycerin, and such was

Dittmar's dualin analysed by me, should surpass, in blasting, a chemically pure Nitro-Glycerin, is to expect 60 cents of currency to have more value than 100 cents of gold, or that a part is greater than the whole.

As I have above referred to my analysis of Mr. Dittmar's dualin, I will give in full the process and result of the same, for the benefit of the reader.

Twenty (20) grammes of dualin were allowed to digest in a glass tube for several days, covered with washed sulphuric ether. The ether was then drawn off, and the residue in the glass tube washed with ether until the cessation of the peculiar persistent taste of Nitro-Glycerin, causing the "Glycerin headache," proved the Nitro-Glycerin was exhausted. The residual woody fibre was now dried thoroughly, and weighed eight grammes. A portion of it thrown on a red-hot plate did not deflagrate; this indicated it had not been treated with nitric acid, and had not been converted into nitro-cellulose. Washed in distilled water, and the washings evaporated, no saline or crystalline salt was obtained. The residue, dried and thrown on a red-hot plate, charred and burnt like any other sawdust. Now, I assert positively, the dualin I analysed, furnished by Mr. Dittmar himself for blasting in the Tunnel, was simply a compound of washed sawdust and Nitro-Glycerin (actually yellow fuming Nitro-Glycerin.)

I have deemed it due to myself to extend these observations further than I intended, but, in the interest of truth, I could not permit the friendly notices of the press, which have been industriously secured, nor the biassed views, of men employed in exploding, (to whom payment of ten dollars was promised, for every case of dualin used, to exaggerate results), to mislead mining contractors, and I stand prepared to prove that 100 parts dualin are only equal to 50 parts pure Nitro-Glycerin, for practical blasting purposes. Dualin is a mixture varying according to the humor of the compounder, but never exceeding one-half the strength of Tri-Nitro-Glycerin; it has all the danger of the Nobel Nitro-Glycerin, with the additional tendency to decomposition, sworn to by Mr. Dittmar himself at the Worcester investigation, owing to its being an admixture of organic matter with Nitro-Glycerin, and its inventor, (as evidenced by his patent, where he proposes to concentrate sulphuric acid, and free it from nitrogen, by boiling it with charcoal!), does not understand the properties of the commonest commercial compounds he undertakes to handle. These facts determine, I submit, the superior advantage of a uniform chemical product produced under invariable conditions, especially since it is more difficult to explode it, and it is proportionately safer, and, above all, has double the effective force.

Mr. Dittmar's promises have failed, and his representations have been disproved by the results at the Hoosac Tunnel. Up to October, 1870, he had six trials, of which he only claims one as a success, though he did succeed in inducing the employees to misrepresent the facts to the contractors, and thereby obtained a testimonial; but over two thousand pounds of his dualin was buried in the Berkshire mountains—a stern pecuniary lesson, verifying the truth of the old Roman apothegm, so much neglected in modern times—"Magna est veritas et prevalebit."

CHAPTER VII.

Nitro-Glycerin Patents and Litigation.

It is seldom that any valuable invention has been brought into public use without costly litigation being entailed on the inventor; and especially is this the case in chemical discoveries, either by pretenders who would interfere with the inventor who has turned his discovery to practical account, on the plea of having previously conceived the same idea, or by unscrupulous individuals who would appropriate to their own use, without payment, the fruits of the labors of other men's brains; hence the writer did not altogether escape, as will be seen by the following remarks on the subject.

I will commence by stating briefly that a patent was granted and four re-issues of the same made to Alfred Nobel and his assignees, for the use of Nitro-Glycerin for blasting purposes,

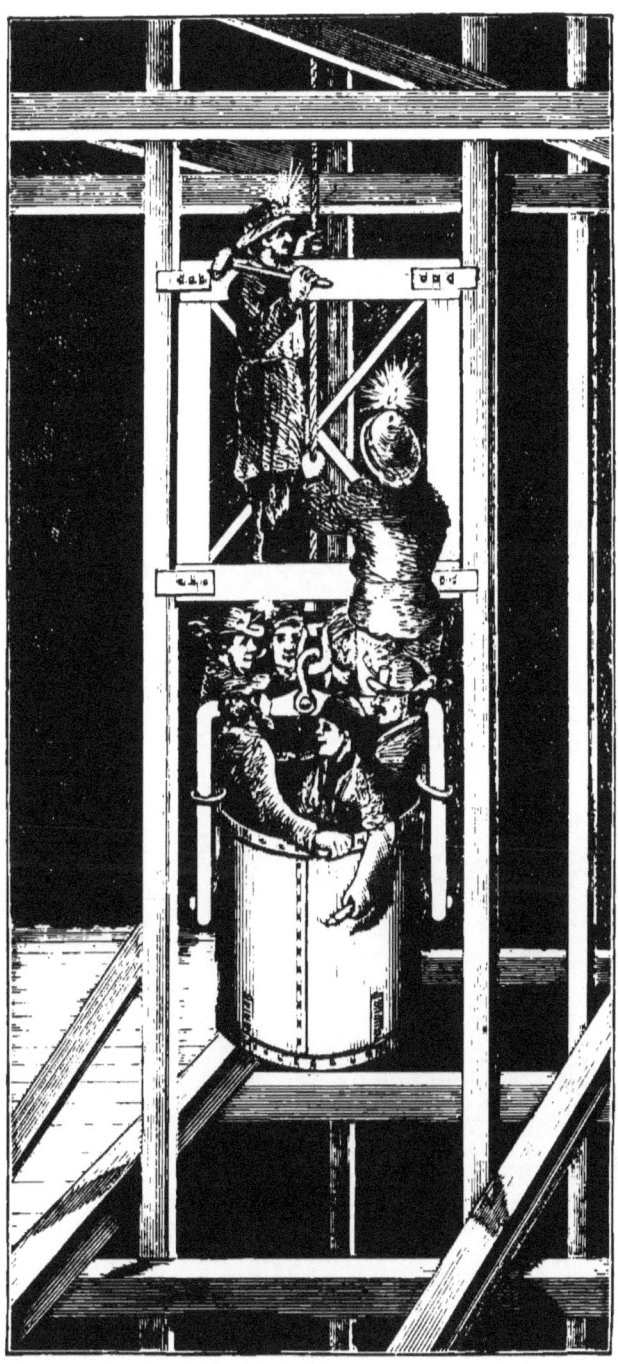

Miners ascending Central Shaft.

when "confined," and for a process of manufacturing the same, by running the glycerin and mixed acids together rapidly, in suitable proportions, into a tank of water. Now, it has never been denied that Sobrero was the discoverer of Nitro-Glycerin, and that it was competent for any one to manufacture that article. The only point, therefore, on which a patent could be obtained was for some improved method of making it. Accordingly, in the course of experiments, I discovered that by passing a current of cold, compressed air through the mixing glycerin and acids, a very valuable improvement was effected, economizing time and material, and rendering the process of manufacturing safer; and for this I obtained a patent on April 7, 1868.

That my readers may see how far I was correct in my estimate of the patentable value of my invention, I give below the opinion of eminent counsel:

NEW YORK, July 10, 1869.

GEO. M. MOWBRAY, ESQ.:

Dear Sir:—Pursuant to your request, I have examined your Letters Patent of the United States for inventions in the manufacture of Nitro-Glycerin, dated the 7th April, 1868. I recollect of aiding you in preparing the application for that patent, and of examining it immediately after it was issued. I believed then that that patent was good and valid, and nothing since has occurred that has changed my opinion or shaken my confidence concerning its validity.

I have recently examined copies of the five re-issued patents to assignees of Alfred Nobel, and I find nothing in them, or any of them, which impairs the validity of your patent.

I further say, that it is my opinion, and clearly so, that the manufacture and sale of Nitro-Glycerin made according to the process described in your patent, does not infringe upon any of the five re-issued patents granted to the assignees of Nobel; and that so far as any of those re-issued patents are concerned, or anything else that I know of, you have a clear right to manufacture and sell Nitro-Glycerin according to your patent.

Very respectfully,

GEO. GIFFORD, Counsellor at Law.

This discovery was not allowed to pass unchallenged, for Mr. Tal. P. Shaffner, having learnt that I had obtained a patent, came forward with a claim that he had conceived the idea (!) in 1865; and in January, 1869, nearly a year after the application for the patent which was granted to me, he applied for a patent for the same thing. This brought our respective rights before the Patent Office in a matter of interference. However, the following remarks by Mr. John W. Thacher, Examiner of Interferences, in giving his decision on the case, will show pretty clearly to whom the right to a patent justly belongs. He says:

"The principle is well established that he who first reduces an invention to practical form is entitled to a patent therefor. Applying this test in this case, the right to a patent seems to rest entirely in Mowbray, and the invention is accordingly awarded to the patentee."

And again Mr. Samuel S. Fisher, the Commissioner of Patents, in giving his decision, remarks:

"The story of Shaffner is not that of a man who had invented anything. He had a theory, talked about it, doubted its value; did not experiment to satisfy himself, until Mowbray was manufacturing on a large scale; and evidently did not intend to apply for a patent at all. I can find none of the ear-marks of a perfected invention, carried beyond the region of experiment; still less of any trace of diligence. Priority is awarded to Mowbray."

As previously noted, the Nobel patent with its re-issues, in four divisions, and twenty-four columns of specifications, containing eight claims drawn up expressly to intercept infringers, specifically, emphatically, and unmistakably insisted:

1st. That Nobel discovered it was necessary to confine Nitro-Glycerin in order to explode it, and that it was practically impossible to explode it unconfined.

2d. That heat and pressure were the agents necessary for a successful explosion of Nitro-Glycerin.

The writer, however, discovered that the heat, pressure and confinement, claimed by the Nobel patent and re-issues, were unnecessary, by charging an open glass tube with Nitro-Glycerin, the glass tube being immersed in water, and the Nitro-Glycerin exploded by the concussion of a cap containing fulminate of mercury, and so succeeded in extricating himself from the domain of the Nobel patents and their particular claims.

But he could not extricate himself from litigation; the insolvent assignee, the United States Blasting Oil Company, clearly perceiving that the monopoly, as they had termed it, was gone, now resorted to the "pis aller" of litigation, misrepresentation, and threatening every one who used Mowbray's Nitro-Glycerin, with the trouble of making affidavits, engaging counsel, and collecting evidence, a by no means to be despised aggressive warfare to contractors, who need all their time, all their capital, and all their ingenuity, to carry out their contracts to a profitable result. Guaranteeing the payment of enforced damages, I met this flank movement by engaging the best counsel, and resolutely set about terminating the pretensions of these patents.

A Suit in Equity was commenced in the Circuit Court of the United States, Western District of Pennsylvania, during the May Term, 1870, by the

UNITED STATES BLASTING OIL COMPANY OF NEW YORK, BY ITS PRESIDENT, TAL. P. SHAFFNER,

vs.

GEO. M MOWBRAY, J. H. KING, CHAS. LOBB, W. L. HOLBROOK, JAMES DICKEY AND A. D. HATFIELD.

As the sworn affidavits in the above case, now pending, are of great importance in substantiating, both practically and legally, the claims urged in previous observations, on behalf of the "Mowbray system" of manufacturing and using Nitro-Glycerin, I give below the substance of the testimony.

Evidence of George F. Barker, Professor of Physiological Chemistry and Toxicology in the Medical Department of Yale College.

"I have carefully examined the several re-issued patents, Nos. 3,377, 3,378, 3,379, 3,380, 3,381 and 3,382, the four former being divisions A, B, C and D, of the re-issued patent, granted upon the surrender of the original patent No. 50,617, dated

October 24th, 1865, and the two latter divisions 1 and 2 of the original patent, also granted to the assignees of Alfred Nobel, on surrender of the original patent No. 57,175, dated August 14th, 1866, granted to said Alfred Nobel. I would further state that in the specifications of the before mentioned re-issues it is asserted that Sobrero discovered that Glycerin was capable of giving, when mixed with sulphuric and nitric acids, a substance analogous to gun cotton, which is true; and that the specifications of the said patents further state that "Sobrero abandoned further research with the declared opinion that its combustion or explosion could not be managed"; which statement, having read all which Sobrero is believed to have published upon the subject, viz.: his papers published in the Comptes Rendus de L'Academie des Sciences, Volume XXIV., page 247, printed in Paris A. D. 1847, and in the Repertoire de Chimie Applique, Volume II., page 400, printed in Paris in 1860, I have entirely failed to find recorded by him as his opinion.

J. E. de Vrij also, in a communication to the British Association, which was read in July, 1851, and is published in the report of the association for the year 1851, page 52 (Notices and Abstracts), states in regard to Nitro-Glycerin, that it "explodes at a moderate heat, as was shown by experiment, detonating when the drops of Nitro-Glycerin on paper were struck a smart blow with a hammer."

The before-mentioned re-issued patents further assert that "in order to explode the whole, or even a large proportion of the mass of Nitro-Glycerin, it is necessary to subject it to confinement or restraint"; which assertion is untrue, for Nitro-Glycerin, when freely exposed to the air in an open vessel or plate, may be and is capable of being readily exploded, without confinement, restraint, or pressure, as I have proved by experiment made at North Adams, on the 17th day of May, 1870, in exploding upon two occasions a quantity of Nitro-Glycerin in an open saucer with great violence, on which occasion the Nitro-Glycerin was exploded by simple concussion in open vessels, the fulminate cap used as the exploder being suspended above the surface of the Nitro-Glycerin in the saucer, and distant nearly two inches from it; so that the application of heat and pressure, or of either of these agencies, is unnecessary.

The said re-issued patents further assert, that "the degree of confinement must be sufficient to allow a pressure upon the Nitro-Glycerin to an extent that 360° F. will be realized, so that decomposition will take place before the liquid can escape the force or heat of the evolved gases of a percussion cap, etc."; whereas I found on the above occasion that when water was interposed between the Nitro-Glycerin and the percussion cap, so that no measurable increase of temperature (much less 360° F.) could possibly occur in the former, the Nitro-Glycerin could be exploded.

In the first experiment three tubes, closed at bottom and containing half an ounce of Nitro-Glycerin each, were placed in water in a tumbler, being supported an inch from the bottom. Into the water in the tumblers, and outside of the tubes, distant from them nearly an inch, the fulminate cap was put. This was then fired, and caused the explosion of the Nitro-Glycerin through the intervening water. In the second experiment, using a tub of water in which eleven tubes containing Nitro-Glycerin were placed, the explosion of six fulminate caps failed to fire the Nitro-Glycerin, the distance from the tubes at which they were placed, nearly or quite ten inches, being too great. In the third experiment five such tubes of Nitro-Glycerin were suspended in a tub of water distant four or five inches from each other, the fulminate cap being inserted in the middle tube. On firing this cap the Nitro-Glycerin in all the tubes was exploded, as judged from the violent effects produced.

The said re-issued patents further assert that "Gun cotton will explode in proportion to the degree of confinement, igniting at 266° F." The celebrated chemist of the English War Department, F. A. Abel, who has made the most extended researches upon gun cotton on record, asserts in his paper published in the Philosophical transactions for 1869 (an abstract of which appears in the Journal of the Chemical Society of London for 1869, Volume XXIII., page 11,) that rows of detached masses of gun cotton, placed on the ground, and extended 4 or 5 feet, have been exploded with most destructive results by firing a small detonating tube in contact with the piece of compressed gun cotton which formed one extremity of the row or train, the explosion of the entire quantity being apparently instantaneous and equally violent throughout." And further that these and similar experiments "appear to indicate decisively that such explosion is not a result of the heat developed by the explosion of the detonating materials."

I have witnessed the manufacture of Nitro-Glycerin as practiced by the defendant Mowbray, at his works situated near the West Shaft of the Hoosac Tunnel, in Massachusetts, and after a full examination of the mode said to have been the invention of Alfred Nobel, and described in the before mentioned re-issued patents, find that the process actually in daily use, at said Mowbray's works, is that described in said Mowbray's patent No. 76,499, dated April 7th, 1868, which process is substantially different from that described in the complainant's re-issues hereinbefore set forth. According to said re-issues, Nobel's process consists in running two separate streams, the one of Glycerin, the other of mixed nitric and sulphuric acids simultaneously into a conical vessel which is perforated at the lower portion thereof, through which perforations the mixture of acids and Glycerin passes into a vessel placed beneath, containing water. In the Mowbray process, a single fine stream of Glycerin is allowed to run into a previously cooled mixture of sulphuric and nitric acids, through and into which cooled mixture of acids is continuously forced, while the Glycerin is entering, a current of atmospheric air, previously artificially dried, compressed and cooled. The action of this current of air is an essentially important and useful one, both upon the process itself and upon the resulting product. First, as to mechanical effects: it thoroughly incorporates the ingredients; it removes in part the nitrous fumes which would otherwise be retained by and contaminate the product, and it cools the mixture by absorbing the heat produced by the chemical reaction of the ingredients. Second, as to the chemical effects: by the action of the oxygen which this air contains it oxidizes the nitrous acid, which may be present in the acids or may be produced in the reaction, to nitric acid, and thus economizes the materials, increases the quantity of the product, and produces a chemically pure article, as is shown by the fact that the Nitro-Glycerin thus produced is perfectly colorless, congeals uniformly at the same degree of temperature and produces, when exploded, no offensive vapors deleterious to the health of the miners using it. Moreover, as, in my opinion, these nitrous fumes tend to induce decomposition in the Nitro-Glycerin and thus to render it unstable, dangerous, and liable to spontaneous explosion, as is demonstrated to be the case in the analogous substance gun cotton, the introduction, in the method of Mowbray, of cold, dry, compressed air into the mixture, in order to get rid of these nitrous fumes, must be regarded as a substantially new invention.

In my opinion, the character of the Nitro-Glycerin is determined by the strength of the acids used in its preparation; the stronger the acids, the purer the product and the more efficient. I verily believe this: first, because it is true of the precisely analogous compound gun cotton, which is prepared in the same way; Hadow having proved, as stated in his paper published in the Quarterly Journal of the Chemical Society of London in 1854, Volume VII., page 201, that at least three products are obtained by acting upon cotton by a mixture of sulphuric and nitric acids, the most explosive being always produced by the strongest acids; and 2nd, because of similar differences observed in Nitro-Glycerin made by different experimenters, and believed by them to be due to like differences in composition; Railton obtained by analysis, as stated in his paper in the Quarterly Journal of the Chemical Society of London for 1854, Volume VII., page 222, the composition now universally adopted as that of Tri-Nitro-Glycerin. De Vrij believes the product he obtained, Journal de Pharmacie, series III., Volume XXVIII., page 38, 1855, to be Tri-Nitro-Glycerin, and Liecke in Dingler's Polytechnisches Journal, Volume CLXXIX., page 157, 1866, gives methods by which Mono-Nitro-Glycerin, Di-Nitro-Glycerin and Tri-Nitro-Glycerin may be produced, the essential difference in these methods being only the strength of the acids employed. Gladstone's Report of the British Association for 1856, page 52 (Notices and Abstracts), has shown that different samples of Nitro-Glycerin differed in properties according to the amount of water contained in the Glycerin; this water, by diluting the acids, making them weaker. Moreover the physiological properties of Nitro-Glycerin have been found by different experiments to differ widely. Sobrero, its discoverer, says a very small quantity taken upon the tongue produces a severe headache for several hours, whence he concludes that it is poisonous. De Vrij in 1851, says that it is not poisonous, and in 1855 that it produces headache, though ten drops caused no symptoms of poisoning in a rabbit. Dr. Herring, in 1849, reported in the American Journal of Science and Arts, series II., Volume VIII., page 257, observed the violent headache produced by $\frac{1}{260}$ of a grain of Nitro-Glycerin or Glonoin, as he proposed to call it, and killed a cat with three drops. Field, in 1858, Pharmaceutical Journal, Volume XVII., page 544, confirmed these results; but Harley and Fuller, reported in the same place, were unable to obtain them by using other specimens of Nitro-Glycerin, though they largely increased the dose. Field consequently says, place given, page 627, "I am daily more convinced of two important facts connected with it, viz.: the great variation in the strength of different specimens, and the very marked difference in the "susceptibility to its influence." In further support of the opinion that several allied but distinct Nitro-Glycerins have been made, the wide difference in density and in congealing point may also be mentioned.

In my opinion the best effect cannot be obtained with commercial acids, owing to their insufficient strength. I have witnessed at the defendant Mowbray's works, at the West shaft of the Hoosac Tunnel, the preparation of the acids used for making the Nitro-Glycerin, commercial acids being found deficient in strength, and in my opinion it is to the use of these stronger acids, combined with the method described in defendant's patent, as above mentioned, that the stability, efficiency, and, above all, the freedom from noxious gases and vapors of the products of combustion of defendant's Nitro-Glycerin is due, when contrasted with that made by complainant, which I have been informed and verily believe is made with acids of commercial strength, and produces, when exploded in a mine, gases and vapors highly deleterious to health.

I have further examined the patent No. 93,113, dated July 27th, 1869, granted to Mowbray, for exploding Nitro-Glycerin, and have experimented with the same, the explosions hereinbefore enumerated having been effected by the method therein described. And this deponent finds that by said Mowbray's process of exploding Nitro-Glycerin, as claimed in his patent, confinement, restraint, or pressure is wholly unnecessary.

In my opinion the same is true in exploding Nitro-Glycerin on a large scale, as I have been informed, and verily believe that upwards of one thousand explosions of Nitro-Glycerin are made weekly in the Hoosac Tunnel by the mode so described in said patent.

I believe, moreover, that the method claimed by Mowbray, in

said patent, differs materially from any of the various modes of exploding Nitro-Glycerin described in the before-mentioned reissues granted to the assignees of A. Nobel, since these various methods specifically require the Nitro-Glycerin to be under confinement, or subjected to heat or pressure when confined, in order to explode it; while Mowbray claims exposing the Nitro-Glycerin to the concussion, agitation, or percussion of a heavy charge, not less than ten or twelve grains of pure fulminate of mercury, which fulminate is fired by passing the electric spark through a priming composition."

June 8, 1870.   GEORGE F. BARKER.

Evidence of S. W. Johnson, Professor of Analytical and Agricultural Chemistry in Yale College.

"I have read the foregoing affidavit of Professor Geo. F. Barker; I witnessed the experiments therein described, and concur in the statement contained in said affidavit."

June 8, 1870.   SAMUEL W. JOHNSON.

Evidence of George M. Mowbray, Operative Chemist.

"About October, 1867, I concluded an agreement with the Commonwealth of Massachusetts, to erect Nitro-Glycerin works near the West Shaft of the Hoosac Tunnel; these erected, I commenced manufacturing Nitro-Glycerin about the 26th day of December, 1867, and with but few intermissions have continued to manufacture it for blasting purposes for the tunnel work ever since. About June 13, 1868, I had a long interview with Mr. Taliaferro P. Shaffner, the complainant in this suit, when the said Shaffner proposed to me a consolidation of interests, and told me, if I would influence J. H. King and Henry Hinckley to advance the sum of seventy-five thousand dollars, that Robert Rennie of the Lodi Chemical Works, of Lodi, New Jersey, would credit him with acids to manufacture Nitro-Glycerin, to the amount of eighty-five thousand dollars, and he would then purchase land about twenty miles up the Hudson river, and manufacture Nitro-Glycerin. The proposal I forwarded to J. H. King and Henry Hinckley, who deemed the same too chimerical to enter upon, more especially since said Shaffner informed me that one-fifth of the consolidated association would have to be paid to one Frederick Smith, one-fifth to said Robert Rennie, and one-fifth to said Shaffner, on behalf of said U. S. Blasting Oil Company's engagements, said Company being deeply indebted to the Lodi Chemical Works, according to the assertion of Joseph Butterworth, the superintendent at Lodi. Mr. Shaffner further informed me that the United States Blasting Oil Company had transferred and assigned all the patent rights conferred by the Nobel patents to him, and he intended to obtain a re-issue of the said patents, and with the individual patents obtained by him, and the patent that had been granted to me in April, 1868, a Company could be formed that would control the supply of Nitro-Glycerin throughout the United States. I soon after consulted with J. H. King and Henry Hinckley, both capitalists, with means, as to the proposals of Tal. P. Shaffner, and the conclusion that we arrived at, was, that, as all the cash capital, and the only practicable method of manufacturing a safe, stable and pure Nitro-Glycerin, was already secured by patent to me, to place seventy-five thousand dollars at the disposal of the parties named by Mr. Shaffner would not be a sensible or prudent course, in view of the condition to which the management of the said Shaffner had reduced the United States Blasting Oil Company's affairs financially, and the failure to supply the demand for Nitro-Glycerin, although the United States Blasting Oil Company had no competitor in New York; so I informed said Shaffner that said Hinckley and King would not advance the money, to wit: seventy-five thousand dollars, under such arrangements, and the proposition fell through. And I would further state, that at each of the various interviews — one of them prolonged for four hours without interruption — the said Tal. P. Shaffner fully admitted to me that any one could or might make Nitro-Glycerin, either by the method described by Sobrero, the inventor, in 1846, or by my patent, granted in 1868, April 7th, without in any way infringing on the patents issued to A. Nobel, and assigned to said Shaffner, as President of the United States Blasting Oil Company. And further, on the 8th December, 1869, I was at Oil City, at the request of the Lake Shore Nitro-Glycerin Works, and assisted in the explosion of one blast in three drill holes of Nitro-Glycerin, using a frictional electric machine, insulated wires, the priming fuse and fulminating charge, as described in Letters Patent, granted to me, July

27th, 1869, and being No. 93,113, and entitled "An Improved Method of Exploding Nitro-Glycerin." I am well informed of the four re-issued patents, Nos. 3,377, 3,378, 3,379 and 3,380, and the methods therein described differ very materially from the method that was practised on the 8th December, 1869, at the Oil City Tunnel, by me, and particularly in this very material respect; whereas, by the method practiced at the Tunnel, an operator can blast simultaneously at will one hundred drill holes; by the methods described in the re-issues above mentioned, it is absolutely impossible to explode two drill holes simultaneously. And this difference between the simultaneous blasting of a number of holes and firing the same number of holes one after the other has been found in actual results to effect an economy of thirty per cent. in the cost of blasting out rock in the Hoosac Tunnel. In a book (Exhibit B), entitled "Liebig and Kopp's annual report of Chemistry for 1847 and 1848", pages 376 and 377, volume 2, published in London in 1850, there is a notice of the comparative power of nitro-cotton and gunpowder, and reference is there made to the nitro-compounds, made from dextrin, glycerin and sugar, as being "similarly explosive preparations," to gun-cotton and nitro-mannite, which latter is described as a cheap substitute for fulminating mercury in the manufacture of percussion caps, and certain comparative experiments with the former (gun-cotton), as to the relative value of the same, compared with gunpowder, are mentioned as having been made by the celebrated powder manufacturers, "Messrs. Hall & Son, of Dartford, in the county of Kent, England." After such publication, the claim made by the said Nobel, or his assignees, in the re-issues before-mentioned, that Nobel discovered that Nitro-Glycerin could be exploded under confinement is invalid, for the fact that Nitro-Glycerin had been described as a similarly explosive preparation to nitro-mannite and nitro-cotton, or gun-cotton, by it sdiscoverer, Sobrero, necessarily involved, and indeed published the circumstance of its only being necessary to subject it to the like conditions of other explosives to effect its explosion. I further state that in four affidavits filed in this Court, on the 25th of February, by Taliaferro P. Shaffner, and T. P. Shaffner and E. A. L. Roberts, jointly, and E. A. L. Roberts singly, and W. M. Shaffner, these parties have sworn that the mode of exploding at the Oil City Tunnel,

December 8th, 1869, was identical and precisely similar to the mode described in a patent granted to said T. P. Shaffner, December 18th, 1868, and re-issued April 13th, 1869, No. 3,375, whilst the very same parties describing the same blasting at said Oil City Tunnel, at the same time, in the same words, and almost word for word throughout, as positively have sworn that it was identical, precisely similar to the mode of blasting described in the re-issues Nos. 3,377, 3,378, 3,379 and 3,380. Neither of these parties were at any time on the ground during the operations therein and thereat (to wit, Oil City Tunnel) performed, except W. M. Shaffner, who was at no time within twenty feet of the parties operating, and who has erroneously stated that water was poured on to the Nitro-Glycerin at the bottom of the hole, which to my certain knowledge was not done. And I ask the attention of this Court, to the affidavits filed in this cause for the plaintiff, and also in a cause of Taliaferro P. Shaffner against the same defendants, filed February 25th, 1870, as completely disproving each other.

February 26, 1870. GEO. M. MOWBRAY.

Evidence of Phillip Mackey and Timothy Lynch, foremen of miners at the Hoosac Tunnel.

"We were employed during the month of September, 1868, at the West Shaft of the Hoosac Tunnel, at the time when Colonel Shaffner, the complainant, was making experiments with Nitro-Glycerin in the said tunnel, and assisted him by drilling holes in the rock to receive the cartridges containing Nitro-Glycerin, and tamping said holes. After the explosion of the said Nitro-Glycerin, we witnessed its effects on the miners. These effects were usually to produce a dryness about the throat, and feeling of thirst, which led the miners to take a drink of water; immediately thereafter the miners would vomit, and such vomiting would be followed by severe headache, rendering it necessary for the miner so affected to be removed to the air, and out of the tunnel, and the effects of such headache would last for from twelve to eighteen hours; in fact, the vapors caused by the Nitro-Glycerin exploded by said Shaffner were of such a noxious character as to disable the miners generally from continuing their work.

"During the past three years we have often examined the Nitro-

Glycerin manufactured by G. M. Mowbray, and been regularly employed as foremen of the miners who drilled the holes for receiving the cartridges of Nitro-Glycerin exploded by said Mowbray and by his assistants, and we declare that Mowbray's Nitro-Glycerin differs greatly in appearance from that used by said Shaffner; that Mowbray's Nitro-Glycerin is colorless almost as water, whereas Shaffner's was orange-colored; that the explosive effects of said Mowbray's Nitro-Glycerin were much greater, so far as we could observe, and that particularly we have noticed the miners do not suffer from any noxious vapors after the firing of blasts of said Mowbray's Nitro-Glycerin, and that during the three years the Nitro-Glycerin made by Mowbray has been used in said Tunnel, there has not been a single case where a miner has been compelled to leave his work by reason of the gases given off by the explosion of Mowbray's Nitro-Glycerin. And we consider that the Nitro-Glycerin made by said Mowbray, and used in the Tunnel; very much safer to handle, and does not give off noxious gases as compared with the Nitro-Glycerin made by the United States Blasting Oil Company of New York, and used by said Shaffner in the Hoosac Tunnel. And we verily believe that if said Nitro-Glycerin were attempted to be used in the Tunnel, now that so general a use is made of Nitro-Glycerin, it would compel the miners to leave their work and seriously retard the progress of the work by reason thereof, for those who could endure it for a time would have to carry out those who are unable to move after inhaling the gases of the Shaffner Nitro-Glycerin, and thus lose time which would otherwise be employed in doing work.

"We consider it utterly useless to confine the Nitro-Glycerin when fired by Mowbray's system."

Feb. 16, 1870.

PHILIP MACKEY,
TIMOTHY LYNCH.

Evidence of John Van Velsor, Superintendent of Mowbray's Nitro-Glycerin works at the Hoosac Tunnel:

"In October, 1868, I was employed to fit up a Nitro-Glycerin factory at Fairport, Ohio, and instruct the hands in the process of manufacturing under Mowbray's patent of April 7th, 1868. I endorse the evidence of Messrs. Mackey and Lynch, as to the difference in appearance and smell between Mowbray's Nitro-

Glycerin and that manufactured under Nobel's patent by the United States Nitro-Glycerin Company.

"I have made under Mowbray's patent upwards of twenty thousand pounds of Nitro-Glycerin, a great portion of which has been exploded in the Hoosac Tunnel, by a method patented by Mr. Mowbray, dated July 27th, 1869, No. 93,113. I have exploded on numerous occasions the Nitro-Glycerin made at said Mowbray's factory, without subjecting the same to confinement, by firing a charge of fulminating mercury, say ten or twelve grains, contained in a wooden or copper cap, by means of the electric spark. I have witnessed the use of Nitro-Glycerin at the West Shaft of the Hoosac Tunnel, both in the bench work and in the heading, where the blasters left the Nitro-Glycerin in the drill-holes entirely unconfined, such being the general practice at the Hoosac Tunnel, so that in case of the wires not conducting the electricity, or in case of the priming being defective and not firing the fulminating charge, the exploder might be removed from the Nitro-Glycerin without danger to the operator. During the eighteen months I have been in the employ of Mr. Mowbray, manufacturing Nitro-Glycerin, he has only made Nitro-Glycerin by his patented method, and by none other.

February 18, 1870. JOHN VAN VELSOR.

Evidence of A. D. Hatfield.

"I have been employed in blasting in the railroad tunnel at Oil City, using Nitro-Glycerin furnished by the Lake Shore Nitro-Glycerin Company, manufactured under Mowbray's patent. In firing and exploding the Nitro-Glycerin I have acted under a license from George M. Mowbray, said Nitro-Glycerin having been exploded without being confined."

February 19, 1870. A. D. HATFIELD.

Evidence of Charles Lobb, Railroad Contractor.

"I have been engaged in tunelling through the hill at Oil City, Pa., for the Jamestown and Franklin Railroad, and have used for that purpose Nitro-Glycerin manufactured by the Lake Shore Nitro-Glycerin Company, under Mowbray's patent of April 7, 1868. I have tried to purchase Nitro-Glycerin from

Tal. P. Shaffner, President of the United States Blasting Oil Company, and have been unable to procure the same. Said Shaffner referred me to E. A. L. Roberts for the purchase of Nitro-Glycerin, and on application to said Roberts was unable to obtain any.

February 19, 1870. CHARLES LOBB.

Evidence of David Crossley.

"I have been engaged in operating oil wells in Pennsylvania, for ten years. On December 6, 1869, I obtained a torpedo containing six pounds of Nitro-Glycerin from the agent of Robert's Torpedo Company, which he said was from New York, and of the best quality. I had it put into an oil well where it was exploded by said agent.

"The explosion of said torpedo, in said well, had the effect of reducing the production of oil in said well from two barrels of oil to one and a half barrels of oil in a day of twenty four hours.

"On the sixteenth day of December, 1869, I put in another torpedo in the same well, which I obtained from the same agent of the same company. It contained the same quantity of Nitro-Glycerin, which was represented to me to be the same as before-mentioned. This torpedo was exploded by the agent in said well on the day last mentioned. Before the explosion of the torpedo in said well, it produced one and a half barrels of oil in a day of twenty-four hours, and the explosion of said torpedo caused no difference in the production of oil from the same well. About the first day of October, 1868, I employed G. M. Mowbray to explode a Nitro-Glycerin torpedo in another well of mine. He exploded said torpedo in said well in my presence. He used in the torpedo six and a quarter pounds of Nitro-Glycerin. The effect of the explosion was to increase the production of said well from five barrels to one hundred barrels in a day of twenty-four hours. After this, Mr. Mowbray put in and exploded other Nitro-Glycerin torpedos in wells for me, and always with the effect of increasing their production.

"Judging from my knowledge as an expert in operating oil wells and the explosion of torpedos of all the various kinds therein, I consider that G. M. Mowbray's Nitro-Glycerin is far more effective than that of any other party, or that his method of exploding is more effective."

February 19, 1870. DAVID CROSSLEY.

Evidence of Jesse Smith, Oil Well Operator.

"In November 1869, I had a torpedo from the Roberts Torpedo Company exploded in my well in Crawford Co., Pa., by their agent. The explosion was an utter failure, one-half the contents of the torpedo still remaining in it; this the agent said was Nitro-Glycerin."

February 19, 1870.   JESSE SMITH.

Evidence of George West.

"I am employed in exploding the Nitro-Glycerin in the holes drilled by the miners in the Oil Creek Tunnel, Pa. I used Nitro-Glycerin from the Lake Shore Nitro-Glycerin Works, which is very different to that of the United States Blasting Oil Company, of New York, and requires a different mode of explosion. I do not use any of the methods described in Nobel's patent of October 24, and re-issued April 13, 1869, for exploding, for the methods therein described would only explode it, if at all, which I doubt, by hazard, and not with certainty, owing to the peculiar properties of the Lake Shore Nitro-Glycerin as compared with what I have seen and used as the Shaffner, or Nobel's Nitro-Glycerin. I endorse all the previous evidence as to the difference between the Nobel or Shaffner Nitro-Glycerin, and that made under Mowbray's patent. The method I have used to explode this Nitro-Glycerin, at the Oil City Tunnel, consists in what is known as the Austrian battery and electric fuse and fulminating shell; that is, an electric machine, whose exciting plate is made of ebonite or hard rubber, with insulated and conducting wire terminals, which are from $\frac{1}{16}$ to $\frac{1}{32}$ of an inch apart, and between those terminal points a priming composition is inserted, through which the electric spark being passed, such priming ignites, giving a flame (insufficient to explode the Nitro-Glycerin, but) sufficient to inflame a fulminating compound, of which there is a heavy charge, and this fulminating compound being exploded by the priming composition, explodes the Nitro-Glycerin. I have never used the method of exploding with gunpowder as described in the Nobel patent, No. 50,617, in the tunnel aforesaid, nor elsewhere, but I have witnessed attempts to explode the Nitro-Glycerin used under Mowbray's Patent by means of fuse and gunpowder, as described by Nobel, where that method failed."

February 19, 1870.   GEORGE WEST.

Sinking the Central Shaft.

Evidence of H. Julius Smith.

"I am engaged in the business of manufacturing electric fuses and introducing explosive compounds to contractors, miners and torpedo men. I have carefully examined the patents in question re-issued to Tal. P. Shaffner, and, I find, by the modes therein described, it is impossible to fire with certainty, and simultaneously, more than two mines charged with Nitro-Glycerin by any of the methods described in said four re-issued patents; and to effect any explosion of Nitro-Glycerin by any of the methods therein described, and materials delivered to the public by the assignees of the inventor Nobel, it is absolutely essential that the Nitro-Glycerin should be confined as described in the reissues in question. I have also carefully examined the patent issued to George M. Mowbray, dated July 27th, 1869, and find that the process therein described of exploding Nitro-Glycerin, does away with the necessity for confining Nitro-Glycerin in order to explode it. I endorse previous evidence from my own experience in regard to exploding Nitro-Glycerin when unconfined under Mowbray's system. I have also manufactured and delivered upward of twenty thousand fuses to the contractors of the Hoosac Tunnel, capable of exploding Nitro-Glycerin when unconfined, at said Hoosac Tunnel. I have been present when the modes described in the re-issues of the Nobel patent have been carefully practised, and entirely failed to fire Nitro-Glycerin, and in one instance immediately after the failure of the Nobel system, I inserted a fuse of the exact description, and with the electric appliances as described in Geo. M. Mowbray's patent, No. 93,113, dated July 27th, 1869, and the result was a successful explosion. The modes described in the Nobel re-issues, Nos. 3,377, 3,378, 3,379 and 3,380, have been abandoned by all parties with whom I am acquainted, who have important works to carry through, requiring Nitro-Glycerin to be exploded, and particularly by the said Tal. P. Shaffner himself, as I have manufactured, sold and delivered to said Shaffner and others, the apparatus and the exploding electrical fuses for firing Nitro-Glycerin made by said Shaffner, and Nitro-Glycerin made by the Lake Shore Nitro-Glycerin Company, which said fuses or electrical exploders, involve a principle of firing Nitro-Glycerin of great practical importance and very recent development, viz., the principle of concussion, so as to effect the explosion of the entire mass of Nitro-Glycerin instantaneously, without requiring the explosion to be transmitted from particle to particle, in this respect differing very materially from the methods described in the Nobel re-issues above referred to, which require, first, confinement, and then heat and pressure, to be developed in the presence of the Nitro-Glycerin.

February 24, 1870. H. JULIUS SMITH.

Evidence of James H. King.

"I am one of the proprietors of the Lake Shore Nitro-Glycerin Works, situated near Painesville, Ohio. I am personally acquainted with Taliaferro P. Shaffner, and endorse all the evidence of G. M. Mowbray as to Shaffner's proposal to consolidate the Nobel and Mowbray patents, and his admission that the parties he represented did not claim the exclusive right to manufacture Nitro-Glycerin. I would state that one W. B. Roberts, of the firm of Roberts & Co., of Titusville, Pennsylvania, informed me that he is one of the Trustees of the United States Blasting Oil Company, and that since the commencement of this suit I have delivered to Roberts & Co., at request of W. B. Roberts, twelve hundred pounds, or thereabouts, of Nitro-Glycerin manufactured by the company of which I am a member.

"I manufacture (as a party interested in the Lake Shore Nitro-Glycerin Works of Painesville) under a license from George M. Mowbray, under a patent to said Mowbray, bearing date April 7th, 1868."

February 25, 1870. J. H. KING.

Evidence of James Dickey.

"I am acquainted with Nobel's system of blasting. I assisted in making ten explosions in Oil City Tunnel, for Charles Lobb, the contractor. We did not use any of the methods of exploding specified in Nobel's or Shaffner's patents. We used the improved electrical machine of H. Julius Smith, patented August 10, 1869, and used the method of firing and fuse described in G. M. Mowbray's patent of July 27, 1869, and which several methods are entirely different from those mentioned in the several patents claimed by complainant in this case. I used in the blasts made by me, the Nitro-Glycerin manufactured by the Lake Shore Nitro-Glycerin Company, under Geo. M. Mowbray's patent, No. 76,499, dated April 7, 1868. I endorse the statements of the miners Mackey and Lynch as to the noxious effects and danger arising from the use of Shaffner's Nitro-Glycerin, and the freedom from the same in that manufactured by Mowbray's system."

February 25, 1870.   JAMES DICKEY.

Evidence of W. S. Holbrook.

"I was engaged along with James Dickey to perform some blasting in Oil Creek Tunnel. I endorse his statement as to the kind of Nitro-Glycerin and the method of exploding used in said tunnel, and further state that we never used any other process or material."

February 25, 1870.   W. S. HOLBROOK.

Evidence of Henry H. Pratt.

"I was foreman at the West Shaft at the Hoosac Tunnel, up to October 15, 1869. In December, 1869, I went to Oil City, Pa., to show Charles Lobb, the contractor for the Jamestown and Franklin Railroad, how to use Nitro-Glycerin for blasting rock. The weather being very cold, warm water was first poured into the holes to prevent the frozen sides of the drilled hole chilling the Nitro-Glycerin. A charge of Nitro-Glycerin was then poured through the water, and a small cartridge of tin being introduced, the charge was fired by means of a frictional electric machine, connected with a priming fuse and a charge of fulminating mercury, being the mode set forth and shewn in the Letters Patent, granted to George M. Mowbray, No. 93,113, and dated July 27th, 1869. I am familiar with the re-issued patents in question, and the mode by which I exploded said Nitro-Glycerin in said tunnel, as above described, is very different from the mode described in the patents re-issued to said U. S. Blasting Oil Company; it would have been utterly impossible to have fired the said three holes in said tunnel by the mode stated in the above referred to re-issues at one and the same moment, as was done by me. I find on examination, that in all the patents granted to Taliaferro P. Shaffner, Nos. 51,671, 51,674, dated December 19th, 1865, the mode of firing a consecutive series of fuses is condemned by said Shaffner, and in patent No. 51,674, that the specification accompanying said Letters Patent contains the following words: "Figures 6 and 7 represent the heretofore known mode of exploding two or more charges by the same electric current, and the former is shewn as applied to a consecutive series of blasts in line, and the latter to the heading of a tunnel," such mode being identically and exactly what I practised at the Oil City tunnel, and none other. I confirm all the previous evidence as to the feasibility of exploding pure Nitro-Glycerin when unconfined, and also as to the good qualities of the Mowbray Nitro-Glycerin when compared with that made under the Nobel re-issues."

February 26, 1870.   H. H. PRATT.

Evidence of Otto Burstenbinder, of New York.

"I have been familiar with the use of Nitro-Glycerin since May, 1865, and introduced that article from Hamburgh, Germany, in July, 1865. I witnessed the application of Nitro-Glycerin to blasting purposes about 20 miles from Hamburgh, when many distinguished citizens were present, a full account of the results effected being published afterwards in the principal German newspapers. The mode used to explode Nitro-Glycerin on that occasion was by fuse and cap, the Nitro-Glycerin being confined, in one experiment, in a gas-pipe, plugged at each end, and the fuse led through the plug, and at the end of the fuse there was a percussion cap attached; in another experiment a wooden plug was hollowed out conically inside and the cone was filled with gunpowder; to this plug a fuse was attached and lighted in the usual manner. I myself fired Nitro-Glycerin in the City of New York, on or about the fifteenth day of July, A. D. 1865; this was the first time I used Nitro-Glycerin in the United States, for blasting purposes; the mode of operation was to pour the Nitro-Glycerin into the naked drill hole, and lower a wooden plug charged with gunpowder, on to the Nitro-Glycerin, poured some dry sand on to the plug, and fire a fuse which was situated on the plug in the usual way.

"I am quite familiar with the Nitro-Glycerin manufactured by the United States Blasting Oil Company, under Nobel's patent, and that manufactured by G. M. Mowbray under his own, and confirm all the previous evidence as to the superiority of Mowbray's Nitro-Glycerin, in explosive power, in absence of color, absence of smell, absence of nitrous gases, in greater safety through the greater difficulty of exploding it, and in purity. As an expert of considerable experience in the use of Nitro-Glycerin, I assert that it is entirely unnecessary to confine Nitro-Glycerin in order to explode the same, the explosion being as thorough, and its effects nearly as powerful for blasting purposes, owing to the extreme instantaneous conversion into gas when unconfined, provided a proper charge of fulminate be used.

"I have made the explosion of Nitro-Glycerin, and its application to blasting purposes, my occupation since 1865, and am thoroughly familiar with its properties, use, and the literature referring to it, and I have never heard or read that the Nitro-Glycerin made by Sobrero was incapable of being crystallized, but I verily believe, and have always found, that Nitro-Glycerin congeals when exposed to a moderately low temperature."

June 7, 1870. OTTO BURSTENBINDER.

Parties using Nitro-Glycerin are requested to note, that on the 19th of March, 1872, the insolvent U. S. Blasting Oil Company (by the aid of funds drawn, under litigation also, from the Oil producers of Pennsylvania, by the notorious torpedo patents), finding their twenty-four columns of specification and eight claims wholly inapplicable to the mode of using Nitro-Glycerin as now practised, surrendered their re-issues, and, as I am of opinion, by the injudicious oversight of the Examiner, an intimate friend of Mr. Shaffner, obtained four more re-issues, containing twenty columns of specification and seventeen claims, thereby, as eminent counsel advise me, practically abandoning their case up to March 19, 1872.

Counsel further advise me, after full consideration of these last re-issues, that the litigation has entered upon a new phase and that the original patent, the first re-issues, and the second re-issues, contain in themselves the proof of their utter worthlessness, needing no other evidence to render them void. But a graver and more serious charge rests upon the means by which these anomalies have been put on record in the Patent Office, which will be reviewed by experienced counsel, before a competent tribunal.

For myself, with resources which I hope and intend to keep unimpaired, to conduct this business to its final issue, with a pecuniary interest I am bound to take care of, besides a further amused interest, aroused during the past four years, by the shifts and pretences of this impecunious company to avoid trial of a suit instituted by itself, there will be a courteous desire to accommodate my opponents with the earliest possible verdict, counsel, judges and jury can arrive at, consistent with a complete, full and fair investigation of plaintiff's pretences and patents.

CHAPTER VIII.

Hoosac Tunnel — Drilling by Machine — Blasting with Powder — Nitro-Glycerin.

The Hoosac Mountain, whose summit is 2,700 feet above the sea level, is composed, according to the geologist, of mica slate, so compressed that near the West End the stratification is contorted, upheaved, and intermingled with quartz and pyrites; consequently the classification of the rock as "mica slate" conveys a very imperfect idea of its hard impracticable nature to the miner. To any one who will be at the pains of examining the masses lying near the powder magazine, built of massive stone, at the West Shaft, the hardness of this rock is at once apparent. Parts of this mountain have been found so hard and tough, and so difficult to drill, that thirty-four drills have been worn in drilling a blast hole thirty-six inches deep. This was an exceptional case, but similar hard layers are met from time to time. Had it not been for the Burleigh drill and Nitro-Glycerin, the sturdy indomitable perseverance of Massachusetts would have been severely strained, if not exhausted, in running this Tunnel.

The following extract from the Adams Transcript, for April 11, 1872, gives a summary of the progress made during the month of March, and the lengths remaining to be opened to complete the work:

Profile of the Hoosac Mountain, and Advance of Tunnel, January 1, 1872.

Drawn By Cap't O. Yäder Kirch

West Shaft 318 Ft To Grade

Jan 1st 1872

West Shaft 850 Ft Above Tide Water

Jan 1st 1872
Central Shaft 1028 Ft Deep To Grade
Jan 1st 1872

Jan 1st 1872

East Shaft 700 Ft Above Tide Water

12194 Ft.

25031 Ft.

12837 Ft.

Section of Brickarch

Section of Rock Excavation.

HOOSAC TUNNEL PROGRESS FOR MARCH, 1872.

"East End, 120 feet; Central Shaft, eastward, 100 feet; West End, 140 feet, total, 360 feet. Total lengths opened to April 1, 1862: East End, 10,166 feet; Central Shaft, east, 617 feet, west, 325 feet, total, 942 feet; West End, 7,494 feet. Lengths remaining to be opened: between East End and Central Shaft, 2,054 feet — 586 feet less than half a mile. Between West End and Central Shaft, 4,375 feet — 855 feet more than two-thirds of a mile."

A reference to the wood cut opposite page 80, shows the profile of the mountain and progress of the Tunnel to January 1, 1872.

The distance made during the month of March, in the East heading, was 120 feet of heading, 24 feet wide and 9 feet in height, exclusive of first enlargement or roof, and second enlargement of roof to full size or stopeing, which is usually carried on simultaneously to about 250 feet per month. This heading is being attacked by twelve of the Burleigh drilling machines, mounted on two carriages manned by eight miners and a foreman, who work for eight hours, with brief intermission whilst the charges are being fired. The drills are impelled by compressed air, making 300 strokes per minute, and calculated to strike with a force of 200 lbs. at each blow, perforating from one inch to five inches per minute, of a hole two inches in diameter when powder is used, and 1¼ inch diameter for Nitro-Glycerin blasting. At the East heading, partly owing to the rock being softer than either at the West End or in the Central Shaft, partly to the miners being accustomed to powder, partly to the heavy battery of drills enabling twelve drilling machines to work at once, and thus make progress satisfactory to the contractors, who, wisely, let well enough alone, the holes when drilled to a depth of from two feet six inches to three feet, are each charged with from one to two and one-half pounds of blasting powder, then tamped; the carriages are drawn back, and the sixteen to twenty-six holes are fired simultaneously by means of a frictional electric machine. This takes place every four hours, exploding from 100 to 150 cartridges every twenty-four hours. The reader must not infer from this that every blast makes from two feet six inches to three feet of advance; because, first, the holes are never drilled for powder in a horizontal plane, but at an angle, sometimes upwards, sometimes downwards, to the right or left, the aim being, that a straight line drawn from the bottom of the hole to the face of the rock shall be shorter than the extreme length of the drilled hole, so that the charge or blast which exerts its force in the line of least resistance, may displace the rock between the bottom of the hole and the surface of the rock, and not collar the hole, that is, merely remove the rock surrounding the outlet of the drilled hole. It is usually found also, that the power exerted by powder is not sufficient, in working a heading, to blast out the rock from the bottom of the hole, but, most frequently, from the point where the cartridge begins, and the tamping terminates. Thus, if a hole be drilled at an acute angle from the face to a depth of thirty inches, with a line of least resistance of twenty-four inches from the bottom of the hole, and a fifteen inch cartridge of blasting powder be inserted, and tamping to the extent of fifteen inches be rammed in above the cartridge, the rock removed, will, under ordinary circumstances, be removed from about where the cartridge commences, that is about 12 inches, or it may be 14 inches, in a direct line from the face. And herein lies the very important distinction between powder and Nitro-Glycerin; the latter, bottoms, i. e., removes the rock from the bottom of (in roofing and quarry work beyond) the hole; with powder this is rarely the case. Moreover, as the depth of the holes is increased, so must the diameter be increased in proportion to the depth when powder is the blasting agent, but when the drilled hole is to be blasted out with Nitro-Glycerin, a diameter of $1\frac{3}{4}$ inches is sufficient for a hole having a depth of ten feet, and a line of least resistance of eight feet, a depth wholly inadmissible for powder, because the rock at that depth would act like the breech of a cannon, and the explosion would issue direct from the hole, only fracturing the edge, i. e., collaring the hole. With Nitro-Glycerin the holes need not be drilled at so acute an angle to the face of the rock, and need no tamping, that is, the drilled hole is left entirely open, and no time is occupied therefore in ramming materials over the explosive, and no risk is incurred in cutting the fuse or electric wire, as with powder, dualin or dynamite, all of which must be tamped. The explosion of Nitro-Glycerin differs from that of every other explosive in this, that the explosion is instantaneous, consequently the rock yields before any flash can reach the mouth of the drilled hole, and the work is done before the gases can travel six feet. Hence the necessity of deep holes; to charge holes only 30 inches deep (except they are from ⅝ to ⅞ inch diameter) is a waste of the material. The same charge will clear the rock to the bottom, with a hole drilled six feet deep, and in fact bottom the six foot hole, whilst a similar charge inserted in a 30 inch hole may leave three or six inches of the hole visible with its surrounding rock, after the blast. And here I cannot refrain from narrating what a narrow escape Nitro-Glycerin had at one time from being rejected at the Tunnel. In the dark days of this enterprise, when every cent expended was narrowly watched, and when it was favor enough for a miner to condescend to allow Nitro-Glycerin to be used in his shift, requests and entreaties for deep holes, and remonstrances that the holes were not drilled deep enough to give this explosive a fair chance, were found fruitless; until, finally, a consultation was held in the time-keeper's office at the West End, the purport of which was, to notify the writer that no more Nitro-Glycerin was needed, as it did not answer expectations. The superintendent, at the West Shaft, was asked what reason I gave that greater progress was not made with the new explosive. His reply was: "Mowbray says the holes are not drilled deep enough, and, I think (he added) it is but fair his demand for deep holes should be complied with, before you throw up the use of Nitro-Glycerin. He has outlaid some $5,000 for the experiment, and you ought at least to see the effect of deep holes, before you decide." Agreed; the superintendent then went to the foreman of the shift, and requested deeper holes, ordering six feet holes. "It's no use," was the reply; "it's all nonsense; why, I tell ye, it won't bottom a hole 30 inches deep; then how is it going to fare with a six foot hole; besides, we can't drill six feet holes by hand in one shift." "Then take two shifts to do it, and take three if it is necessary; this Nitro-Glycerin man says he must have deep holes, and he shall for this once, if I drill them myself, and it takes a week to do it."

The deep (only six feet) holes were drilled, and charged; cartridges of same size as those inserted in 30 inch holes, were used, and fired, every hole bottomed, every miner was astonished, and from that day the use of Nitro-Glycerin was a necessity for the heading in the West End. But it was a narrow escape from what would have been deemed a failure. On another occasion, during a drought, the supply of water at the West End, where the Nitro-Glycerin was manufactured, gave out, and, being a necessity in the manufacture, we had to haul it by team. This was troublesome work, and cost money. There had been a change of engineers, and the gentleman now in charge, on the difficulty reaching him, determined first to ascertain whether Nitro-Glycerin was a necessity, before complying with the contract the Commissioners had made, and which involved a supply of compressed air and water, if they used Nitro-Glycerin. And to make no mistake, the holes of what is termed the "cut" in the heading, that is, two series of four holes each, in a parallel line from the roof, about nine feet high, were drilled about five feet apart at the face of the heading, and six feet deep, tending towards each other so that at the bottom of the holes they terminated about three feet apart. After charging and firing, the above gentleman and his assistant inspected the result. A mass of rock eight feet in height, five feet wide in front, and about five feet deep, with the rear end three feet wide, had been blown from its seat, some ten feet from the heading, and there stood, a monument (until block-holed) of the use of Nitro-Glycerin, when properly applied. "You shall have all the water you want, sir, if I bring it myself in pails," was the energetic assurance of this gentleman, who felt satisfied that Nitro-Glycerin was a necessity for the Hoosac Tunnel.

In drilling holes for blasting with Nitro-Glycerin, a depth of not less than five feet should be reached; six feet are better, but ten and twelve feet are the right depth for a heading, whilst fifteen feet for bench work, and eight feet apart, or, for quarry work ten feet apart, and ten feet from the face, provided the rock is hard enough (in clay, owing to the sudden shock Nitro-Glycerin is ineffective); exploded in holes of such a depth it will throw out everything before it — and make progress. How difficult to get miners to drill such holes, how many frivolous objections, how the wires and their connections will be tampered with to interfere with the intended blast, and how criminal, contrary, and pig-headed, they deem the contractor and Nitro-Glycerin man who insists on such depth of holes, I have often

"Stopeing out" Roof Enlargement (East End.)

experienced, and it needs the firmness and vim of desperation to enter a quarry, descend a shaft, or go into a rock cutting, and oppose the life-long habits of men who believe honestly they know everything that concerns mining, and what they do not know is not worth knowing. But if once a blast is shewn, and they have to hoist out the rock, their obstinacy succumbs, and in three months, men, who knew it was poison, and so dangerous it was wicked to ask them to drill holes to receive it, have positively refused to descend a shaft if powder was attempted to be used merely in a comparative experiment, alleging, that the powder was unhealthy and not fit to be used at the bottom of a shaft, where the air was confined. And here let me truly add, I have never sent Nitro-Glycerin to be experimented with in any rock work, rock cutting, or rock tunnel, that was not followed by a large order, repeated until the end of the work, during my past experience of four years' manufacture. Indeed, there have been only two cases where it was found inapplicable, and these were in hard clay, where it seems actually to mould for itself a chamber, compressing the walls of the drill hole, as if an enormous hydraulic ram had been inserted; but the tenacious mass is not displaced, it only suffers compression. When, therefore, holes can be made with a crow-bar, and not drilled, do not use Nitro-Glycerin, but if you have rock, be it as hard as emery, or as the magnetic iron ore of the Lake Superior or Ottawa Iron mines, the harder the better for the economy of drilling, which is very great, so few holes being required, the introduction of Nitro-Glycerin, with a good steam or air drill, causes the progress of the work to advance to that degree that it is only limited by the ability to remove the debris of blasted material. To return from this digression to my subject.

To effect this progress of 120 feet, probably about 3,000 holes have been drilled in an area not exceeding 24 feet by ten feet, requiring twelve drilling machines, and 60 horse steam power to compress the air requisite to drive the drills; add to this the powder, over a ton and a half, the electric exploders, the candles and oil for miners, and the fact that a mass of rock 120 feet long, ten feet high and twenty-four feet wide, has to be carried out and dumped two miles from where it was excavated, and some slight idea of the labor at this one point may be formed. Now take double this length of rock, viz.: 250 feet, increase its height to 15 feet, keeping its breadth of 24 feet — I say, take this mass which is torn from the roof, whilst the heading is being pushed, and bring it and dump it 1¾ miles from where it lay solid, and you have again another point on which you can begin to estimate the East End work. About 350 men, a locomotive, forty cars, 200 horse water power, machinists, blacksmiths a legion, for sharpening drills is hand work, so is picking up rock, loading cars, making track, and all this is done in the smoky, wet, grimy, confined tunnel, or round about its entrance, and you have a mixed, confused suspicion that this tunnel driving is a work needing high powers of organization; and, with the license of the miner, his pay day, his weddings and his wakes and funerals, which are all powerful reasons for quitting work, you have a still clearer idea of the anxiety such work involves

CENTRAL SHAFT.

The Plate, opposite page 74, conveys an idea of the sinking of the Central Shaft at 891 feet depth; at the time of writing, May, 1872, however, this shaft had not only reached grade, but to a sump beneath grade at a depth of 1,040 feet; headings and enlargements have been also driven at grade, east and west, to meet the works from the East End, and from the Western Shaft. Owing to the stratification of the rock, which dips towards the west, great progress was anticipated in this direction; but man proposes and God disposes; on reaching about 300 feet westward, seams of water were struck, of so threatening a nature that a powerful Cornish pump was erected, at a cost reaching, in all its details, $80,000, and now, May, after enlarging the diameter of the former plunger pump, prudence suggests the temporary delay of any further disturbance of this water inlet (immediately under the divide of the mountain), until the present pumping force has sufficiently drained the sources of water supply to permit a further advance of this (the western) heading of the Central Shaft to be driven without involving a flooding out of the men working at the eastern heading. Meanwhile, from the sump, the excavations are enlarged to full tunnel size, the capacity of the Cornish and plunger pumps are being tested, and all energy summoned to meet any difficulties to be overcome when this western heading of the Central Shaft shall resume work. All the rock here has to be moved from the heading by hand power, and lifted (by steam power) 1,000 feet to the surface, yet, notwithstanding these adverse circumstances, during March, 100 feet was driven to the eastward alone. I append a memorandum furnished by Mr. E. A. Bond, of actual drilling and blasting, taken at this point during the dates given, being about the average performance.

On August 19th, 1871, on the north side of the east heading, machine No. 1, starting at 10 A. M., had at 2.08 P. M. drilled three holes, averaging about five feet four inches; the time actually occupied in drilling being 74 minutes, or an average of about 25 minutes to each hole. The remaining 2 hours and 54 minutes are accounted for by changes of drills, breaking of carriage, and an interval of 40 minutes for dinner. On the south side, machine No. 2, starting at 9.35 A. M., had at 2.09 P. M. drilled three holes, averaging about six feet four inches; the time actually occupied in drilling being 81 minutes, or an average of 27 minutes to each hole. The remaining 3 hours and 13 minutes are accounted for in a similar manner to the time of machine No. 1, except that there was no accident to the carriage. The average time of the two machines was about 26 minutes for the average depth of about five feet ten inches, being two inches and seven-tenths per minute. It will be seen by these facts that the actual drilling is but a comparatively small part of the work; bringing forward the machines, connecting to the air main, inserting the drills into the jaws of the machine piston, changing these drills as they wear down, oiling, releasing drill when stuck, removing back the machine carriage out of reach of the blasted rock, waiting for blaster to charge the holes, connect his wires, and apply the electric current to fire the exploders, removing the debris to clear the track for the approach of the drills — all these operations, so varied and yet so necessary, each consume a considerable quota of the eight hours allotted to each shift.

On August 30, 1871, a blast was made in the east heading at 5.30 P. M., as follows: fourteen 7 foot holes were fired with 25 lbs. of Nitro-Glycerin, throwing out about 30 tons of loose rock; and one solid rock, diameter 9 x 4½ x 4 feet, and weighing about 24,000 lbs., a distance of 30 feet, a weighty testimonial to the explosive power of Nitro-Glycerin.

The expense incurred and difficulties met with, in working at the Central Shaft, will serve as a hint to contractors to make all due allowance in their estimates for striking a seam of water; work may go on smoothly for a long time; the general geological formation of hill or mountain may be well understood, and yet the contractor cannot tell but that he may strike a vein of quartz that may throw him back days and weeks in his drilling calculations, or a seam of water which will cost him thousands of dollars in machinery and labor to keep it under.

On December 7, 1870, the hoisting machinery broke at the Central Shaft, and then the following measurements of water were made. On December 3, the depth was 3 feet; December 13, 7 feet; December 15, $8\frac{1}{2}$ feet; December 20, $21\frac{1}{8}$ feet; and December 24, $48\frac{1}{2}$ feet. At midnight they commenced bailing with two buckets, one having a capacity of 341 gallons or 54.65 cubic feet, and the other $189\frac{1}{2}$ gallons or 31.36 cubic feet. The large bucket was hoisted 1,075 times, bailing 58,745.3 cubic feet of water, and the small bucket 966 times, with 29,327.8 cubic feet of water, the whole amount being 549,179.0 gallons in 27 days, or 21,080.0 gallons per day.

The following anecdote is worth relating, as showing the wonderful escapes men sometimes have, when the chances are one hundred thousand to one against their lives:

In February, 1872, Thomas Hawkins felt tired and sleepy, and concluded to lie down in the east heading of the Central Shaft, about 30 feet distant from where the blaster was charging sixteen holes with Nitro-Glycerin, intending to retire when the holes were charged. But he failed, as we many of us do, to carry out his intention. When the blaster had charged his holes, he left the heading, connected his wires, and having hallooed the usual warning "Fire," and every thing being quiet, discharged his blast. Thomas Hawkins was awakened by the report of the blast, scattering 30 or 40 tons of rock, and annoyed to find his foot bruised, he limped out to meet the miners returning to their work, who now, when a blast is about to take place, unceasingly ask him where he proposes to take up his position, that they may choose an equally safe place.

An escape, as wonderful, at the West Shaft, is worthy of being recorded. On August 3, 1868, as Richard Dunn was advancing to the heading, with a can about a quarter filled with Nitro-Glycerin, his foot slipped, and, in trying to avoid falling, he swung the can over his head, striking the drilling-machine frame, and fell prostrate, still holding the can; a rush of air was heard, and the can was found as shown in the photograph, page 66, the Nitro-Glycerin not having exploded. The man got up a great deal more unconcerned than those at work near him, and quietly went forward and filled his cartridges as if nothing had happened. As I told him afterwards, he will never be so near eternity again without actually reaching it. The can had been filled at a temperature of 45° F., and the temperature of the room where it had been stored for 36 hours, was about 65°, thus causing an expansion both of the Nitro-Glycerin and the air contained in the can.

The West End of the Tunnel comprises the brick arch and portal, well No. 4, the supplementary shaft, and what is known as the West Shaft. The brick arch has been driven through what is aptly termed, "demoralized rock," for immediately after the spring thaw it becomes a quicksand, and spews into the tunnel from every direction. By driving small adits on each side, and a central adit some distance ahead of the main tunnel, Mr. B. H Farren overcame this dangerous and difficult work, which at one time threatened his contract, and thus enabled the arch work to be carried on. Subsequently, the central adit was carried through to the West Shaft, and thus the costly and difficult task of lifting 420 gallons of water per minute, to a height of 320 feet, was avoided, and it now escapes by natural flow through the west portal. Drilling is practised here as described for the East End and Central Shaft; in the East End the heading is driven on grade, and the overhanging enlargement is "stoped" out by hand drilling worked from an arched stage, (see plate opposite page 85) thus avoiding the necessity of handling twice; mules draw the laden trucks, from the heading and beyond where this stopeing out of the roof is going on, to the locomotive, which hauls a train of cars laden with stone to the dump.

At the West End, however, the roof of the heading is driven in line with the roof of the tunnel, which is hereby left complete as the heading progresses; this involves trucking by hand, and dumping the rock from the heading over the bench to the lower level, see plate opposite page 90, and is not found so economical as the East End method. These differing methods of working, however, were not started simply as experiments, but for good engineering reasons; at the East End, the dump was ample below the grade of the outlet, whereas, at the West End there was no opportunity to get out at the portal, on the line of the intended railroad; all the rock here had to be lifted (until the portal and arched work were completed) up and out of the West Shaft, and dumped on to the mountain side, and, to avoid being impeded by water, the heading was driven on a level higher than the grade of the Tunnel, thus ensuring good drainage for the most important part of the work, as it was then deemed, viz.: monthly linear advance. For the Commissioners were servants of the public, and the advance, rather than the enlargement of the Tunnel, was the measure of their success so far as public opinion was concerned.

Only by a personal visit to this enormous work can a correct idea be obtained of the expense, ingenuity, engineering skill, and indomitable energy of the several foremen and superintendents at the four divisions, viz.: East End, under Mr. Blue; at the Central Shaft, under Mr. Roskrow; at the West Shaft, Mr. Williams, with underground superintendent, Mr. White; and at the West Portal or arch work, the sub-contractors, Messrs. Hocking and Holbrook; all of whom are daily devising more expeditious methods of detail, in compassing the great end sought by each brigade, the completion of the Hoosac Tunnel contract at the time specified.

And whilst this energy, this organization, and all this development of the highest grade of modern engineering, are being devoted to carrying out the expressed wish of the majority of the people of Massachusetts, the malcontent minority is sleepless in offering every possible obstruction to the work; in Governor's council, in consulting engineering supervision, in committee of assembly, in the newspaper press, covert expression of the opposition has found vent, and been doubtless useful in its way. But is it not time this opposition should cease? Must our citizens be for ever confined to one route from their Capitol to the West? Surely there will be traffic enough and ample, to remunerate both lines, when the Hoosac Tunnel route is open. If so, the time is approaching for a generous welcome from the opponents of the Hoosac Tunnel, and the conditions "at owner's risk and at corporation's convenience" may cease to appear on our freight notes.

Instructions for Handling and Using

MOWBRAY'S

TRI-NITRO-GLYCERIN.

---

1. Handle carefully, avoiding a sudden jar or concussion, and be very careful, if any is spilt outside the can, to avoid striking it against any hard substance.

2. When solid, thaw out by placing the cans in a tub of warm water, not hotter than the wrist can bear, first pouring warm water into the can, and always remove the can before adding more hot water to the tub.

3. To fill Cartridges, &c. — Hold the Cartridges to be filled over a tray, say 2 feet by 3 feet, the bottom of which should be covered with Plaster of Paris (which will not readily explode when saturated with Nitro-Glycerin.) The soiled Plaster of Paris should be frequently renewed.

4. If the Nitro-Glycerin in a liquid state is kept in store or magazine for some time, the cork should be loosely inserted, and a pint of cold water poured in each can, to be frequently poured off and replaced with fresh cold water in warm weather, taking care to retain the bladder under the cork. It is preferable, when ice can be procured, to congeal the Nitro-Glycerin.

5. Use Funnels (gutta-percha if they can be had) for filling water holes. Under no circumstances whatever attempt to tamp the drill holes; it is unnecessary, and may kill the man who attempts it.

6. Hot irons to warm the water, or soldering the cans, will be sure to cause explosions.

7. Never sledge or attempt drilling in a hole or seam where Nitro-Glycerin has been spilled; fire an exploder, which will effectually clear it up.

8. Never pour Nitro-Glycerin into a hole unless perfectly sure that it is a sound hole, or will hold water; if seamy always use cartridges.

9. To obtain the best results with Nitro-Glycerin, drill deep holes, 6 feet or more. Use powerful exploders and well insulated wires. It is cheaper to fire by electric battery with simultaneous explosion, than to fire several holes with tape fuse.

10. Look out after a blast for any unexploded cartridges lying around.

11. Never allow any but the most careful persons to handle or have charge of the Nitro-Glycerin, and insist upon the use of every precaution to prevent an accident or explosion.

12. Never allow empty Glycerin cans to be used for any other purpose, but destroy them by a fuse and exploder, or building a fire under them, first, however, removing them to a safe distance.

13. Examine your cans from time to time, and notice if, at the level of the Nitro-Glycerin, any pin-holes have eaten through; in such case procure a new can, or stone jar, and empty the contents out, not trusting your hold to the upper part of the can, lest it may give way.

14. When solid, or congealed, it is absolutely safe; if possible, therefore, any surplus should be stored surrounded with ice, since no explosion can take place when it is solid.

GEORGE M. MOWBRAY.

North Adams, Mass., June, 1872.

APPENDIX.

A.

MEMORANDA FOR CONTRACTORS.

1. There are very different qualities of Nitro-Glycerin, varying from 50 per cent. in blasting force, and the same manufacturer, unless able to control absolutely every detail of his work, cannot insure a precisely similar product, even from similar ingredients.

2. The best Nitro-Glycerin may be simply fired, or only exploded, or its full blasting effects achieved, precisely according to the initial velocity or force used to start the explosion; two cents in an exploder therefore may save ten dollars in a blast.

3. Ten per cent. of water diffused through Nitro-Glycerin, giving it a milky appearance (Nitro-Glycerin emulsion), will diminish its effective blasting results 30 per cent.

4. Thirty per cent. more blasting power is evolved, when the Nitro-Glycerin reaches the bare rock of the drill hole, than when, by insertion in cartridge, the metal of the cartridge and a layer of air or water are interposed between the blasting gases and the rock.

5. Pure Nitro-Glycerin may be safely stored, and does not readily change; impure Nitro-Glycerin needs only time and temperature to explode spontaneously.

6. In hard pan, or indurated clay, Nitro-Glycerin is not so economical as powder; in granite, gneiss, hornblende, quartz and other hard rocks, the harder the better, especially in large erratic boulders, the larger the better, Nitro-Glycerin will enable the tunneling, cut or block-holing, to be performed at half the cost as compared with gunpowder.

B.

"OVER-SENSITIVE" EXPLODERS.

The term, "over-sensitive," has been used in the fore-going pages, and applied to exploders. Mr. Joseph Dowse, of Lockport, Illinois, applied "fulminate of copper" (a discovery of Dr. John Davy) as a priming for exploders, and patented the application, observing in his patent that parties unaccustomed to the preparation of fulminates had better leave this preparation alone. The sequel shows Mr. Dowse's caution was not superfluous. Two manufacturers, provoked by the commercial inconvenience of the constant return of exploders owing to their inefficiency, have resorted to this "over-sensitive" priming, and received the following warnings:

In 1869, Mr. Stowell was standing in the office, on Sudbury street, Boston, whilst Mr. H. Julius Smith was packing 200 exploders in a rubber bag, in which an ebonite electric machine had been placed. Mr. Stowell remarked, "Is it safe to crowd them into a bag like that?" "Oh yes, perfectly safe," was the reply, when instantly 170 out of the 200 exploded, severely burning and injuring both Smith and Stowell, the latter being confined to his bed for five weeks in consequence.

A similar explosion occurred to Mr. Smith on another occasion, the copper caps penetrating the fleshy part of the thigh, in almost the same parts as Mr. Stowell had been wounded, and burning the eyelashes, eyebrows and face severely; by this accident Mr. Smith was confined to his room for a considerable time.

Mr. Smith's partner, in touching some of this priming, whilst moist, in a wooden bowl, was also severely burnt by its detonation, the face, eyebrows and eyelashes being injured, and himself confined to his room for four days.

On Thanksgiving day, 1869, Charles A. Brown was handling some of this priming, incautiously touching it on a piece of glass with a steel knife; it exploded, and the consequence has been deprivation of sight.

One Hogan, in the Fall of 1871, working in Charles A. Brown's exploder factory, lost the sight of one eye, the other being severely injured, by imprudently omitting his helmet (usually worn whilst handling this material), and proceeding to move some of the primers whilst drying the same.

The superintendent, foreman of machine shop, foreman carpenter and blaster, engaged in connecting the wires, at the enlargement of the East End, were killed April 21, 1871, by a premature explosion, caused by the lightning striking the iron rails, whence the induced and ambient electricity, radiating to the leading wire, fired the over-sensitive exploders which were inserted in the charges of Nitro-Glycerin.

At the Burleigh Mine, Georgetown, two men were killed from similar causes producing similar effects.

An exploder, from one of the above manufacturers, placed in a cartridge that was being lowered with forty pounds of Nitro-Glycerin from the Government scow, at Dimon's reef, to the diver below, exploded by reason of the friction of the insulating wire as it passed through the hands of Superintendent Pierce; now, as there were 300 pounds of Nitro-Glycerin on the scow, had it exploded, it must have destroyed the scow and every soul (about 40) on board. Fortunately, the fulminating charge was as imperfect as the priming was over-sensitive, confirming remarks on page 42.

These casualties, the comments of the press, together with the constant explosions in the factories of those who prepare "over-sensitive" exploders, are beginning to influence both principals and employees, and it is hoped exploder makers will eventually succeed in either resorting to the Abel priming, or discover, in the records of the Patent office, some formula that they can imitate, not so sensitive as that of Mr. Jacob Dowse, and whose proprietor is equally indifferent, or not "over-sensitive" to infringement. It is too much to expect they will surprise their friends, as Sheridan is reported to have astonished his, when, after repeated failures to guess how he became possessed of a new pair of boots, he coolly announced, "he had actually bought and paid for them."

Meanwhile, the manufacturer of Nitro-Glycerin, if he would avoid the additional risk of exploder accidents, which are invariably laid to Nitro-Glycerin, must make his own exploders, and try to construct the necessary electric apparatus to fire them, until further developments have stimulated those who have entered into these trades to perfect their wares.

C.

PROFESSOR ABEL ON EFFECTS OF INITIAL EXPLOSION ON EXPLOSIVES.

Mr. Abel, of the Woolwich Arsenal, Great Britain, in an abstract of the Proc. Royal Society xvi. 395, observes :

The degree of rapidity with which an explosive substance undergoes metamorphosis, as also the nature and results of such change, are in the greater number of instances susceptible of several modifications, by variation of the circumstances under which the conditions essential to chemical change are fulfilled. Excellent illustrations of the modes by which such modifications may be brought about are furnished by gun-cotton, which may be made to burn very slowly and almost without flame, to inflame with great rapidity, but without development of great explosive force, or to exercise a violent destructive action; according as the mode of applying heat, the circumstances attending its application, and the mechanical conditions of the explosive agent are modified. Nitro-Glycerin or Glonoin, which bears some resemblance to chloride of nitrogen in the suddenness of its explosion, requires the fulfillment of special conditions for the full development of its explosive force. Its explosion by the simple action of heat can be accomplished only when the source of heat is applied for a considerable time in such a way that chemical decomposition is established in some portion of the mass, and is favored by the continued application of heat to that part; under these circumstances the chemical change proceeds with very rapidly accelerating violence, and eventually brings about a sudden transformation of the heated portion into gaseous products, which transformation is instantly communicated throughout the mass of Nitro-Glycerin, so that confinement of the substance is not necessary to develop its full explosive force. This result can be obtained more expeditiously, and with greater certainty, by exposing the substance to the concussive action of a detonation produced by the ignition of a small quantity of fulminating powder placed in contact with or near to the Nitro-Glycerin.

The development of the violent explosive action of Nitro-Glycerin, freely exposed to air, through the agency of a detonation, was regarded until recently as a peculiarity of that substance; but Abel's experiments have shown that gun-cotton and other explosive compounds and mixtures do not necessarily require confinement for the full development of their explosive force; this result being obtained (and very readily in some instances, especially in that of gun-cotton) by means similar to those applied in the case of Nitro-Glycerin, viz.: by the percussive action of a detonation.

The action of a detonation in determining the violent explosion of gun-cotton, Nitro-Glycerin, etc., cannot be ascribed to the direct operation of the heat developed by the chemical changes of the charge of detonating compound used as the exploding agent. An experimental comparison of the mechanical force exerted by different explosive compounds, and by the same compound employed in different ways, has shown that the remarkable power exhibited by the explosion of small quantities of certain bodies (the mercuric and argentic fulminates) to accomplish the detonation of gun-cotton, while comparatively large quantities of other highly explosive agents are incapable of producing this result, is generally accounted for in a satisfactory manner by the difference in the amount of force suddenly brought to bear in the different instances upon some portion of the mass operated upon. Most generally, therefore, the degree of facility with which the detonation of a substance will develop similar changes in a neighboring explosive substance may be regarded as proportionate to the amount of force developed within the shortest space of time by that detonation, the latter being, in fact, analogous in its operation to that of a blow from a hammer, or of the impact of a projectile. Several remarkable results of an exceptional character have, however, been obtained, which indicate that the development of explosive force under the circumstances referred to, is not always simply ascribable to the sudden operation of mechanical force. Thus silver fulminate, which explodes much more suddenly, and with much more powerful local force than mercuric fulminate, nevertheless, when applied under the same conditions, does not induce the explosion of gun-cotton so readily as mercuric fulminate. Five grains of mercuric fulminate enclosed in a case of stout sheet metal, and exploded in close contact with compressed gun-cotton, caused the detonation of the latter, but five grains of silver fulminate enclosed in tin-foil, though it appeared to produce quite as sharp a detonation as the same quantity of the mercury salt enclosed in the stout case, did not explode the gun-cotton with which it was surrounded, but merely scattered the mass; when enclosed in the stout sheet metal case, however, the five grains of silver fulminate accomplished the detonation of the gun-cotton. Iodide and chloride of nitrogen are much more susceptible of sudden explosion even than silver fulminate; nevertheless, the iodide does not appear to be capable of causing the explosion of compressed gun-cotton; and the chloride of nitrogen shows but little capability of producing the same effect, fifty grains being the smallest quantity that will answer the purpose.

Lastly, it is found that Nitro-Glycerin when exploded by a charge of mercuric fulminate, will not bring about the explosion of compressed gun-cotton placed in contact with it, though under precisely similar circumstances the explosion of gun-cotton or of Nitro-Glycerin will induce the explosion of a larger mass of its own kind.

These results point to the conclusion, that the effect of the detonation of one substance in causing the explosion of another depends not only on the force, but also on the nature of the vibrations developed in the former; the most probable explanation of the observed results being that the vibrations attendant upon a particular explosion, if synchronous with those which would result from the explosion of a neighbouring substance in a state of high chemical tension, will, by their tendency to develop those vibrations, either determine the explosion, or, at least, greatly aid the disturbing effect of mechanical force suddenly applied, while, in the instance of another explosion, which develops vibratory impulses of a different character, the mechanical force applied through its agency, has to operate with little or no aid, so that greater force or a more powerful detonation is required in the latter case to accomplish the same result.

D.

NITRO-GLYCERIN CAR OFF THE TRACK.

The perfect safety with which Nitro-Glycerin can be transported, when congealed, is demonstrated in the following fact, which should effectually banish from the minds of freight agents and express companies the objections which they have heretofore successfully urged against carrying Nitro-Glycerin by rail; so far, at least, as concerns that manufactured by the writer.

On May 3, 1872, a special car loaded with seventy-nine cans containing 4,800 pounds of congealed Nitro-Glycerin, was being transported over the Chesapeake and Ohio Railroad, from Huntington to Charlestown; C. J. Cheshire, Assisting-Superintendent at the Maysville, Ky., Works, was on the car running at the rate of 18 miles an hour; suddenly the car jumped the track, and was dragged over the ties, some of which were two feet ten inches measured distance apart (the new roadway not then ballasted), for a distance of 684 feet, before the train could be brought to a stand still, to the no small consternation of Mr. Cheshire, the engine-driver and stoker. The rough jolting had no effect whatever on the Nitro-Glycerin, except tumbling some of the cans off the car, and in a few hours, the car being replaced, transportation was resumed, and one more experience of the properties of our Nitro-Glycerin added to the list.

E.

ACCIDENTS AT THE HOOSAC TUNNEL.

Until within the last two years there has been no complete record kept in the State Engineer's office of the casualties among the miners at work on this great undertaking; but a careful examination of the existing records, and of the superintendents at different portions of the work, has enabled us to present the following analysis of the accidents, causing death or injuries to miners, which have occurred within the past three years, and to this we append the accidents by gun-cotton, Erhardt's powder and fire, which, although of an earlier date, from their peculiar nature have had special memoranda made in regard to them.

ANALYSIS.

	Killed.	Injured.
Killed and injured by falling rocks, tumbling down Shaft, and the usual casualties of miners other than those mentioned below,	14	12
Fire — Burning Central Shaft,	13	
Over-sensitive Exploders,	7	a number.
Dualin (about 600 lbs. actually used),	1	3
Erhardt's Powder (less than 500 lbs. used),	3	10
Gun-Cotton (about 250 lbs. used),	1	4
Nitro-Glycerin (about 150,000 lbs. used),	5	5
Gun-Powder (most of the accidents from powder, occurred at an earlier date than our record, which in this respect is necessarily incomplete),	2	3
	46	37
		8
		45

This analysis shows 46 killed, and 45 (allowing 8 as the "number" vaguely mentioned in the records) injured by the various sources of accidents referred to, and as the relation of Nitro-Glycerin to other explosives is what especially interests our readers, the following comparative analysis of the deaths in proportion to the number of pounds of each explosive used at the Hoosac Tunnel, will enable them to form some idea as to the comparative safety of those mentioned.

ANALYSIS.

	Killed.	Amount used. lbs.	Proportion of deaths per 100 lbs.
Erhardt's Powder,	3	500	.6
Gun-Cotton,	1	250	.4
Dualin,	1	600	.16
Nitro-Glycerin,	5	150,000	.0003

As Nitro-Glycerin has 13 times the explosive power of gunpowder, our readers, who are accustomed to use the latter for blasting, can easily ascertain the percentage of accidents in proportion to the amount used, and so judge for themselves as to the comparative safety of these explosives.

Really, whilst using, only two lives have been lost; one man rashly advancing to the charge, although advised to desist, whilst his fuse was burning; the other, on change of shift, after a blast, a cartridge having failed to explode, and the blaster neglecting to examine whether his cartridge had exploded, allowed the new shift to proceed drilling in the same rock, and within one inch of the same spot previously drilled, and where a charged cartridge was contained, when after a few inches of drilling progress, they came on to the concealed cartridge — explosion followed. In the magazine where three were killed, in order to hurry up, after a previous night's spree, it had become the practice, notwithstanding peremptory warnings, to remove the cover of the stove, and expose the naked can of Nitro-Glycerin to the naked fire, of course, explosion must, as it did, follow this reprehensible folly, and disobedience to orders, resulting in killing three men.

I have established Tri-Nitro-Glycerin Factories

At North Adams, Massachusetts,
ALFRED WALLACE, Foreman;

At Maysville, Kentucky,
JOHN WALLACE, Superintendent;

At Kingston, Province Ontario, Upper Canada,
H. H. PRATT, Superintendent;

In order to facilitate supply, and make deliveries at least possible cost for freight.

GEO. M. MOWBRAY,
NORTH ADAMS, MASS.

Where orders for Exploders, both electric and tape fuse, gutta-percha insulated leading and connecting wire, of quality very superior to any hitherto made in the United States, should be addressed.

Agent in New York City:

W. B. TOWNSEND,

No. 40 Broadway (Room 39.)

www.ingramcontent.com/pod-product-compliance
Lightning Source LLC
Chambersburg PA
CBHW020121170426
43199CB00009B/588